T0221637

Advanced Sustainable Forest Management

Advanced Sustainable Forest Management

Edited by **Aduardo Hapke**

New York

Published by Callisto Reference,
106 Park Avenue, Suite 200,
New York, NY 10016, USA
www.callistoreference.com

Advanced Sustainable Forest Management
Edited by Aduardo Hapke

© 2015 Callisto Reference

International Standard Book Number: 978-1-63239-030-1 (Hardback)

This book contains information obtained from authentic and highly regarded sources. Copyright for all individual chapters remain with the respective authors as indicated. A wide variety of references are listed. Permission and sources are indicated; for detailed attributions, please refer to the permissions page. Reasonable efforts have been made to publish reliable data and information, but the authors, editors and publisher cannot assume any responsibility for the validity of all materials or the consequences of their use.

The publisher's policy is to use permanent paper from mills that operate a sustainable forestry policy. Furthermore, the publisher ensures that the text paper and cover boards used have met acceptable environmental accreditation standards.

Trademark Notice: Registered trademark of products or corporate names are used only for explanation and identification without intent to infringe.

Printed in the United States of America.

Contents

Preface   VII

Section 1   Introduction   1

Chapter 1   **Sustainable Forest Management:**
An Introduction and Overview   3
Jorge Martín-García and Julio Javier Diez

Section 2   **Carbon and Forest Resources**   17

Chapter 2   **The Quality of Detailed Land**
Cover Maps in Highly Bio-Diverse Areas:
Lessons Learned from the Mexican Experience   19
Stéphane Couturier

Chapter 3   **Remote Monitoring for Forest**
Management in the Brazilian Amazon   45
André Monteiro and Carlos Souza Jr.

Chapter 4   **Sustainable Management of Lenga (***Nothofagus pumilio***)**
Forests Through Group Selection System   65
Pablo M. López Bernal, Guillermo E. Defossé,
Pamela C. Quinteros and José O. Bava

Chapter 5   **Case Study of the Effects of the Japanese Verified**
Emissions Reduction (J-VER) System on Joint Forest
Production of Timber and Carbon Sequestration   87
Tohru Nakajima

Section 3   **Forest Health**   109

Chapter 6   **Cambial Cell Production and Structure of Xylem and**
Phloem as an Indicator of Tree Vitality: A Review   111
Jožica Gričar

Chapter 7   **A Common-Pool Resource Approach to Forest Health:**
             The Case of the Southern Pine Beetle   135
             John Schelhas and Joseph Molnar

Chapter 8   **Evaluating Abiotic Factors Related to Forest Diseases:**
             Tool for Sustainable Forest Management   149
             Ludmila La Manna

Section 4   **Protective and Productive Functions**   165

Chapter 9   **Soil Compaction – Impact of Harvesters'**
             and Forwarders' Passages on Plant Growth   167
             Roman Gebauer, Jindřich Neruda, Radomír Ulrich
             and Milena Martinková

Chapter 10   **Ecological Consequences of Increased Biomass**
             Removal for Bioenergy from Boreal Forests   185
             Nicholas Clarke

Section 5   **Biological Diversity**   197

Chapter 11   **Ecological and Environmental Role of**
             Deadwood in Managed and Unmanaged Forests   199
             Alessandro Paletto, Fabrizio Ferretti, Isabella De Meo,
             Paolo Cantiani and Marco Focacci

Chapter 12   **Close-to-Nature Forest Management:**
             The Danish Approach to Sustainable Forestry   219
             Jørgen Bo Larsen

             Permissions

             List of Contributors

Preface

It is often said that books are a boon to mankind. They document every progress and pass on the knowledge from one generation to the other. They play a crucial role in our lives. Thus I was both excited and nervous while editing this book. I was pleased by the thought of being able to make a mark but I was also nervous to do it right because the future of students depends upon it. Hence, I took a few months to research further into the discipline, revise my knowledge and also explore some more aspects. Post this process, I begun with the editing of this book.

This book covers advances made in Sustainable Forest Management (SFM). Although it is an old concept, its popularity has grown in the last few decades due to the public concern regarding the significant decline in forest resources. The implementation of SFM is usually achieved with the help of criteria and indicators (C&I). A number of countries have developed their own sets of C&Is. The book covers some of the current researches carried out to test the recent indicators, to look for new indicators and to establish new markers. It contains original research studies on biodiversity, carbon & forest resources, forest health and productive & protective functions. These studies highlight the recent researches done to provide forest managers with effective tools for choosing between distinct management strategies of improving indicators of SFM.

I thank my publisher with all my heart for considering me worthy of this unparalleled opportunity and for showing unwavering faith in my skills. I would also like to thank the editorial team who worked closely with me at every step and contributed immensely towards the successful completion of this book. Last but not the least, I wish to thank my friends and colleagues for their support.

Editor

Section 1

Introduction

Sustainable Forest Management: An Introduction and Overview

Jorge Martín-García[1, 2] and Julio Javier Diez[1]

[1]Sustainable Forest Management Research Institute,
University of Valladolid – INIA, Palencia
[2]Forestry Engineering, University of Extremadura, Plasencia
Spain

1. Introduction

It is well known that forests provide both tangible and intangible benefits. These benefits may be classified according to ecological values (climate stabilization, soil enrichment and protection, regulation of water cycles, improved biodiversity, purification of air, CO_2 sinks, potential source of new products for the pharmaceutical industry, etc.), social values (recreational and leisure area, tradition uses, landscape, employment, etc) and economic values (timber, non wood forest products, employment, etc.). Although forests have traditionally been managed by society, it is expected that the current growth in the world population (now > 7,000 million people) and the high economic growth of developing countries will lead to greater use of natural resources and of forest resources in particular.

2. Global forest resources

The total forest area worldwide, previously estimated at 4 billion hectares, has decreased alarmingly in the last few decades, although the rate of deforestation and loss of forest from natural causes has slowed down from 16 million hectares per year in the 1990s to around 13 million hectares per year in the last decade (FAO, 2011). Nevertheless, the loss of forest varies according to the region, and while the forest area in North America, Europe and Asia has increased in the past two decades (1990-2010), it has decreased in other regions such as Africa and Central and South America, and to a lesser extent Oceania (Fig. 1)

There is growing public concern about the importance of the environment and its protection, as manifested by the fact that the total area of forest within protected systems has increased by 94 million hectares in the past two decades, reaching 13% of all the world's forests. Moreover, designated areas for conservation of biological diversity and for protection of soil and water account for 12 and 8% of the world's forests, respectively (FAO, 2010, 2011). Nevertheless, other statistics such as the disturbing decrease in primary forests[1] (40 million hectares in the last decade) and the increase in planted forests (up to 7% of the

---

[1] Forest of native species where there are no clearly visible indications of human activities and the ecological processes have not been significantly disturbed (FAO, 2010)

world's forests) (FAO, 2011) appear to indicate that to achieve forest sustainability, we must
go beyond analysis of the changes in the total forest area worldwide.

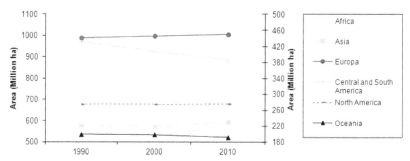

Fig. 1. State of World's Forests 2011 – subregional breakdown (Source: FAO, 2011). Africa,
Asia, Europe, Central and South America and North America are represented in the left axis
and Oceania in the right axis.

3. Sustainable forest management

The concept of sustainability began to increase in importance at the end of the 1980s and at
the beginning of the 1990s with the Brundtland report (1987) and the Conference on
Environment and Development held in Rio de Janeiro, Brazil, in 1992 (the so-called Earth
Summit), respectively. Nevertheless, the need to preserve natural resources for use by future
generations had long been recognised.

The negative influence of past use of forest resources, as well as the needs for continued use
of these resources for future generations was already noted as early as the 17th century
(Glacken, 1976, as cited in Wiersum, 1995). However, it was not until the 18th century that
the concept of sustainability was specifically referred to, as follows: "every wise forest
director has to have evaluated the forest stands without losing time, to utilize them to the
greatest possible extent, but still in a way that future generations will have at least as much
benefit as the living generation" (Schmutzenhofer, 1992, as cited in Wiersum, 1995). This
first definition was based on the principle of sustainable forest yield, with the main goal
being sustained timber production, and it was assumed that if stands that are suitable for
timber production are sustained, then non wood forest products will also be sustained (Peng
2000). This assumption focused on the sustainability of the productive functions of forest
resources, while other functions such as ecological or socio-economic functions were largely
overlooked. This occurred because social demands for forests were mainly utilitarian.
However, increased environmental awareness and improved scientific knowledge
regarding deterioration of the environment have changed society's values and the global
structural policy, which in turn have significantly influenced forest management objectives
in 20th century (Wang & Wilson, 2007). Nevertheless, nowadays more and more researchers
think climate change is changing the paradigm and sustainability shouldn't be referred to
what we had before.

Although there is no universally accepted definition of SFM, the following concepts are
widely accepted: "*the process of managing permanent forest land to achieve one or more clearly
specified objectives of management with regard to the production of a continuous flow of desired
forest products and services without undue reduction of its inherent values and future productivity*

and without undue undesirable effects on the physical and social environment" (proposed by International Tropical Timber Organization: ITTO, 1992), and *"the stewardship and use of forests and forest lands in a way, and at a rate, that maintains their biodiversity, productivity, regeneration capacity, vitality and their potential to fulfill, now and in the future, relevant ecological, economic and social functions, at local, national, and global levels, and that does not cause damage to other ecosystems"* (proposed by the second ministerial conference for the protection of the forest: MCPFE, 1993). The latter concept harmonizes ecological and socio-economic concerns at different scales of management and for different time periods. Nevertheless, both concepts are just refining the definition of sustainable development gave by the Brundtland Commission (1987) "development that meets the needs of the present without compromising the ability of future generations to meet their own needs" to apply it to forests.

4. Criteria and indicators

The implementation of SFM is generally achieved using criteria and indicators (C&I). Criteria are categories of conditions or processes whereby sustainable forest management can be assessed, whereas quantitative indicators are chosen to provide measurable features of the criteria and can be monitored periodically to detect trends (Brand, 1997; Wijewardana, 2008) and qualitative indicators are developed to describe the overall policies, institutions and instruments regarding SFM (Forest Europe, 2011).

Different studies have pointed out the main characteristics of a good indicator. Thus, Prabhu et al. (2001) suggested seven attributes to improve the quality of indicators (precision of definition, diagnostic specificity, sensitivity to change or stress, ease of detection, recording and interpretation, ability to summarize or integrate information, reliability and appeal to users), whereas Dale & Beyeler (2001) established eight prerequisites to selection (ease of measurement, sensitivity to stresses on the system, responsive to stress in a predictable manner, anticipatory, able to predict changes that can be averted by management actions, integrative, known response to disturbances, anthropogenic stresses and changes over time, and low variability in response).

Although several criticisms have been launched against the C&I system (Bass, 2001; Gough et al., 2008; Poore, 2003; Prabhu et al., 2001), the popularity of the system is evident from the effort invested in its development in recent decades and from the large number of countries that are implementing their own sets of C&I within the framework of the nine international or regional process (African Timber Organization [ATO], Dry Forest in Asia, Dry Zone Africa, International Tropical Timber Organization [ITTO], Lepaterique of Central America, Montreal Process, Near East, Pan-European Forest [also known as the Ministerial Conference on the Protection of Forest in Europe, MCPFE] and Tarapoto of the Amazon Forest). Nevertheless, three of these processes stand out against the others[2], namely the ITTO, MCPFE and Montreal processes. The first set of C&I was developed by ITTO (1992) for sustainable management of tropical forest, and subsequently an initiative to develop C&I for sustainable management of boreal and temperate forests took place in Canada, under the supervision of the Conference on Security and Cooperation, in 1993. This first initiative reached a general consensus about the guidelines that should be

---

[2] Together, these three international C&I processes represent countries where more than 90% of the world's temperate and boreal forests, and 80% of the world's tropical forests are located.

followed by all participating countries. It was then decided that the countries should be split into two groups: European would establish the MCPFE and non-European countries the Montreal processes. The MCPFE process adopted a first draft of C&I in the first expert level follow-up meeting in Geneva in June 1994, which took shape in Resolution L2 adopted at the third Ministerial Conference on the Protection of Forest in Europe held in Lisbon (MCPFE, 1998), and improved at the subsequent Ministerial Conference held in Vienna (MCPFE, 2003). On the other hand, the Montreal process established its set of C&I in the Santiago Agreement (1995), with Criteria 1-6 improved at the 18th meeting in Buenos Aires, Argentina (TAC, 2007) and criterion 7 improved at the 20th meeting in Jeju, Republic of Korea (TAC, 2009).

Although the different processes have very different origins and have developed their own criteria, there are some similarities between the three major SFM programs (Table 1). The main difference concerns criterion 7, developed by the Montreal process (Legal, policy and institutional framework), which was imbedded within each of the criteria in the MCPFE process (McDonald & Lane, 2004) and the concept of which is similar to criterion 1 in the ITTO process (Enabling condition). One important difference between ITTO and the other two processes is that the former does not consider maintenance of the forest contribution to global carbon cycles.

ITTO process	MCPFE process	Montreal process
C1. Enabling condition	C1. Maintenance and appropriate enhancement of forest resources and their contribution to global carbon cycles	C1. Conservation of biological diversity
C2. Extent and condition of forests	C2. Maintenance of forest ecosystem health and vitality	C2. Maintenance of productive capacity of forest ecosystems
C3. Forest ecosystem health	C3. Maintenance and encouragement of productive functions of forests (wood and non-wood)	C3. Maintenance of forest ecosystem health and vitality
C4. Forest production	C4. Maintenance, conservation and appropriate enhancement of biological diversity in forest ecosystems	C4. Conservation and maintenance of soil and water resources
C5. Biological diversity	C5. Maintenance and appropriate enhancement of protective functions in forest management (notably soil and water)	C5. Maintenance of forest contribution to global carbon cycles
C6. Soil and water protection	C6. Maintenance of other socioeconomic functions and conditions	C6. Maintenance and enhancement of long-term multiple socio-economic benefits to meet the needs of societies
C7. Economic, social and cultural aspects		C7. Legal, policy and institutional framework

Table 1. Criteria for sustainable forest management: comparison of three major programs

Other differences in indicators developed by the different processes have become apparent, and e.g. Hickey & Innes (2008) established more than 2000 separate indicators using the context analysis method. There are also substantial differences as regards the three major processes: the MCPFE process has 52 indicators (MCPFE, 2003), whereas the Montreal process has reduced the number of indicators from 67 (Santiago Agreement, 1995) to 54 (TAC, 2009), and the ITTO process has reduced the number of indicators from 66 in the first revision (ITTO, 1998) to the 56 considered at present (ITTO, 2005).

In light of the proliferation of C&I processes, the need to achieve harmonization has been widely recognised (Brand, 1997; Castañeda, 2000). Although the concept of harmonization is subject to several interpretations, harmonization should not be mistaken for standardization (Rametsteiner, 2006). Köhl et al (2000) has claimed that "harmonization should be based on existing concepts which should be brought together in a way to be more easy to compare, which could be seen as a bottom up approach starting from an existing divergence and ending in a state of comparability". Although there is not yet a common approach, considerable efforts have been made since the first expert meeting on the harmonization of Criteria and Indicators for SFM, held in Rome in 1995 (FAO, 1995), towards the search for a harmonization/collaboration among C&I processes through the Inter-Criteria and Indicator Process Collaboration Workshop (USDA, 2009). Advances in harmonization will minimise costs (avoiding duplication and preventing overlap), facilitate comparisons between countries and, overall, improve the credibility of SFM.

Although indicators are increasingly used, their utility is still controversial. Some authors have pointed out several weaknesses of the indicators, e.g. that they are often highly idealistic (Bass, 2001; Michalos, 1997), that they are a pathological corruption of the reductionist approach to science (Bradbury, 1996) or even that the same indicator may lead to contradictory conclusions according to the criterion and the scale. Nevertheless, there is general agreement that the advantages of the approach outweigh these limitations and that researchers should focus their efforts on testing the current indicators and searching for new indicators.

There are two key aspects involved in improving the current and future indicators, the use of a suitable scale and the establishment of a specific interpretation of each indicator. Although these have mainly been implemented at a national level, sub-national and forest management unit (FMU) levels are essential to assess SFM (Wijewardana, 2008). The FMU level has been considered as the finest scale in C&I processes. However it is well-known that for some indicators (mainly biodiversity indicators), another subdivision within this level may be necessary, such as plot, landscape and spatial levels, for correct interpretation (Barbaro et al., 2007; Heikkinen et al., 2004). In light of this level of precision and the fact that values of indicators are sometimes correlated with several different scales, managers and researchers should establish the most effective scale in each case, to avoid additional charges. Moreover, good indicators are not always easy to interpret in terms of sustainability, because most indicators do not exhibit a clear distinction/threshold between sustainability and unsustainability. In such cases, the achievement of sustainability should be considered on the basis of relative improvement in the current status of the indicator in question (Bertrand et al., 2008).

On the other hand, the scientific community must search for new indicators. Gaps in knowledge have been identified, and as these mainly involve ecological aspects, researchers should go further in investigating the relationships between type of forest management and ecological and socioeconomic functions. Thus, managers and researchers, with the support of scientific knowledge and public consultations, should be able to determine feasible goals, from socioeconomic and scientific points of view, since goals that are too pretentious may lead to a situation whereby SFM will not be promoted (Michalos, 1997). Only then can successful selection of new indicators of SFM be achieved.

5. Forest certification

In addition to the efforts of different states to develop C&I in the last two decades, a parallel process has been developed to promote SFM. This process is termed "forest certification". Forest certification can be defined by a voluntary system conducted by a qualified and independent third party who verifies that forest management is based on a predetermined standard and identifies the products with a label. The standard is based on the C&I approach and the label, which can be identified by the consumer, is used to identify products. Therefore, the two main objectives of forest certification are to improve forest management (reaching SFM) and to ensure market access for certified products (Gafo et al., 2011).

The first certification was carried out in Indonesia in 1990 by the SmartWood programme of the Rainforest Alliance (Crossley, 1995, as cited in Elliot, 2000). However forest certification became popular after The Earth Summit in Rio de Janeiro in 1992. Although important advances were reached at this summit, the failure to sign a global convention on forestry led environmental and non-governmental organizations to establish private systems of governance to promote SFM. In 1993, an initiative led by environmental groups, foresters and timber companies resulted in creation of the Forest Stewardship Council (FSC). Subsequently, other initiatives at international and national levels gave rise to many other schemes, e.g. the Programme for the Endorsement of Forest Certification (PEFC, previously termed Pan European Forest Certification), the Canadian Standards Association (CSA), the Sustainable Forestry Initiative (SFI), the Chile Forest Certification Corporation (CERTFOR) and the Malaysian Timber Certification Council, among others.

The area of certified forest increased rapidly in the 1990s and from then on more gradually, reaching 375 million hectares in May 2011 (UNENCE/FAO, 2011), which represents almost 10% of the global forest area. Although many forest certification systems were developed in the 1990s, only two schemes (PEFC and FSC) have been used for most of the forest currently certified throughout the world. The FSC scheme was established in 1993 to close the gap identified after the Earth Summit, and with more than 140 million hectares is the first program in terms of number of certified countries (81 countries) and the second system in terms of certified area at the moment (FSC, 2011). The PEFC scheme was established in 1999 as an alternative to the FSC scheme, and was led by European forest owners, who considered that FSC standards mainly applied to large tropical forests, but were inappropriate for small forest owners of European temperate forests. The PEFC scheme has gained importance because it endorses 30 national forest certification systems (Australian Forestry Standard, CSA, SFI, CERTFOR, etc.), and with more than 230 million hectares of certified forests is currently the largest forest certification system (PEFC, 2011). Although several authors have reported significant differences between FSC and PEFC (Clark & Kozar, 2011; Rotherham, 2011; Sprang, 2001), detailed analysis has revealed that FSC and PEFC are highly compatible, despite having arrived at their C&I by different routes (ITS Global, 2011).

Although forest certification began in tropical forests, the trend has changed and the scheme is now carried out in boreal and temperate forests. Almost 90% of forests certified by the two major programs (FSC and PEFC) are located within Europe and North America (Figure 2). More than half (54%) of the forests in Europe (excluding the Russian Federation) have already been certified, and almost one third of the forest area in North America has been certified (Figure 3). On the contrary, only about 1.5% of the forests in Africa, Asia, and Central and South America have been certified (Figure 3), despite the fact that more of half of the world's forests and almost 60% of primary world forests are located in these countries. The FSC and PEFC schemes display similar patterns of certification, since both mainly certify forests in Europe and North America. However, although the percentage of forest area certified by FSC in Africa, Asia, and Central and South America is only 16% of all certifications carried out by this scheme, this represents 75% of the forest areas certified in these regions. Furthermore, almost all certifications carried out in the Russian Federation are carried out by the FSC, whereas the PEFC has certified very few forests in this region. On the other hand, most forest certifications in Europe (excluding the Russian Federation) and North America have been carried out by PEFC (Figure 2).

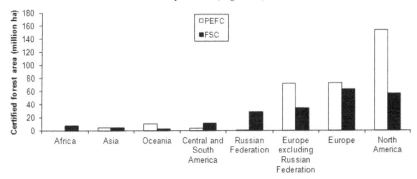

Fig. 2. Global FSC and PEFC certified forest area November 2011 – subregional breakdown (Source: FSC, 2011; PEFC, 2011)

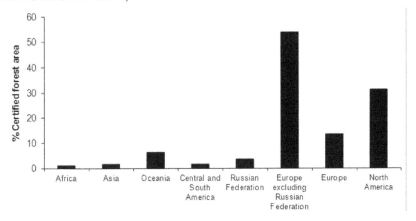

Fig. 3. Percentage of certified forest area, by both FSC and PEFC schemes, November 2011 – subregional breakdown (Source: FSC, 2011; PEFC, 2011)

Forest certification has became very popular, mainly because it is regarded it as a tool whereby everyone should benefit (win-win situation): forest owners should have an exclusive market with premium prices, the forest industry should improve its green corporate image, should not be held responsible for deforestation, and should have available a market tool, consumers should be able to use forest products with a clear conscience, and overall, forests should be managed sustainably.

The concept of forest certification is based on an economic balance, where forest owners and the forest industry place sustainable products on the market in the hope that consumers will be willing to pay the extra cost implied by SFM. Nevertheless, forest certification is still far from reaching its initial goal (win-win), since the expected price increases have not occurred (Cubbage et al., 2010; Gafo et al., 2011). In practice, only consumers and the forest industry have benefited; consumers use certified forest products with a clear conscience, and the forest industry has ensured market access without any great extra cost because this has mainly been assumed by forest owners.

This leads to a difficult question, namely, are forests benefiting from forest certification? It appears logical to believe that forest certification is beneficial to forests, since forest owners must demonstrate that the forests are being managed sustainably. Nevertheless, in depth-analysis reveals a different picture. As already noted, forest certification began in tropical forests with the aim of decreasing deforestation. However, nowadays almost all certified forests are located in developed countries. Furthermore, most of these forests are productive forests, such as single-species and even-aged forests or plantations, in which only small changes must be made to achieve forest certification, while primary forests have largely been ignored. The fact that foresters are able to place certified products from productive forests on the market, with a small additional charge compared to the extra charge involved in certifying products from primary forests hinders certification of the latter, which are actually the most endangered forests. Moreover, this disadvantage may favour unsustainable management, such as illegal logging or in extreme cases conversion of forest land to agricultural land, to favour market competitiveness. Against this background, other initiatives beyond of forest certification has been implemented, such as the FLEGT (Forest Law Enforcement, Governance and Trade) Action Plan of the European Union that provides a number of measures to exclude illegal timber from markets, to improve the supply of legal timber and to increase the demand for wood coming from responsibly managed forests (www.euflegt.efi.int) or the REDD (Reducing Emissions from Deforestation and Forest Degradation) initiative of the United Nations to create a financial value for the carbon stored in forests, offering incentives for developing countries to reduce emissions from forested lands and invest in low-carbon paths to sustainable development, including the role of conservation, sustainable management of forests and enhancement of forest carbon stocks (www.un-redd.org).

In addition, some environmental organizations now consider that plantations should not be certified, since they consider that plantations are not real forests. Such organizations also denounce the replacement of primary forests with plantations in developing countries (WRM, 2010). Although the replacement of primary forests with plantations is a damaging process, replacement of degraded areas such as abandoned pasture or agricultural land provides obvious advantages from economic and ecological points of view (Brockerhoff et al., 2008; Carnus et al., 2006; Hartley, 2002). The two most important schemes (FSC and PECF) approve the certification of forest plantations because they believe that the promotion of wood products from plantations will help to reduce the pressure on primary forests. The

FSC has added another principle (Principle 10: Plantations) in an attempt to ensure SFM in plantations, while the PECF considers that its criteria and indicators are sufficient to ensure the sustainability of planted forests. The FORSEE project was carried out in order to test the suitability of MCPFE indicators (which are used as the basis for PEFC certification in Europe) for planted forests at a regional level in eight Atlantic regions of Europe (Tomé & Farrell, 2009). This project concluded that with few exceptions, the MCPFE criteria and indicators appear suited to assess the sustainable management of forests, although it was pointed out that they should be considered as a blueprint for true SFM and adaptations are needed at the local level (Martres et al., 2011).

The viability of tropical forest certification will depend on forest owners obtaining premium prices that at least cover the certification costs, taking into account that these costs vary according to the type of forest (primary forest, plantations, etc.) and that consumers' willingness to pay premium prices will also differ. It should be possible for consumers to distinguish the origin of each product, and in other words different labels are required. Nevertheless, the use of different eco-labels is controversial, since many labels may confuse rather than help consumers. Teisl et al (2002) noted that consumers "seem to prefer information presented in a standardized format so that they can compare the environmental features between products" and highlighted "the need for education efforts to both publicize and inform consumers about how to use and interpret the eco-labels". Both of these are difficult tasks when different certifiers are rivals in the market place.

Without standardization and a powerful information campaign, most environmentally concerned consumers will probably demand wood from sustainably managed forests, without taking into account the type of certification label, and will choose the least expensive product (Teisl et al., 2002). This may entail a new associated problem, since producers and industries will probably also choose the bodies that certify forests most readily and at the lowest cost. This may lead to a situation where the certification schemes would tend to compete with each other and standards would be reduced to attract producers, as pointed out by Van Dam (2001).

6. Conclusion

Sustainable forest management is evolving with public awareness and scientific knowledge, and the sustainability concept must be revised to reflect the new reality generated by climate change, where a past reference point shouldn't be considered. Therefore, C&I should be updated continuously to be able to cope with the climate change challenge and assess sustainability of changing ecosystems. Furthermore, harmonization of C&I processes would be the most desirable outcome, since this would improve the credibility of the schemes.

On the other hand, forest certification has failed to avoid deforestation and has got two main challenges;

(1) to certify the forests that are most important in ecological terms and that are most susceptible to poor forest management, such as tropical forests and, to a lesser extent, non productive forest in boreal and temperate regions, and (2) to achieve a market with premium prices, in which the win-win concept will prevail. This will require educational campaigns and a higher level of credibility for labels. Moreover, parallel initiatives, such as FLEG and REDD, considering outside forest sector drivers leading to deforestation should be taking into account to limit this process.

7. Acknowledgment

The authors thank Christophe Orazio for helpful comments on earlier versions of the manuscript.

8. References

Barbaro, L.; Rossi, J-P.; Vetillard, F.; Nezan, J. & Jactel, H. (2007). The spatial distribution of birds and carabid beetles in pine plantation forests: the role of landscape composition and structure. Journal of Biogeography, Vol. 34, pp. 652-664.

Bass, S. (2001). Policy inflation, capacity constraints: can criteria and indicators bridge the gap? In: *Criteria and Indicators of Sustainable Forest Management*, R.J. Raison; A.G. Brown & D.W. Flinn (Eds.), 19-37. IUFRO research series, Vol. 7. CABI Publishing, Oxford.

Bertrand, N.; Jones, L.; Hasler, B.; Omodei-Zorini, L.; Petit, S. & Contini, C. (2008). Limits and targets for a regional sustainability of assessment: an interdisciplinary exploration of the threshold concept. In: Sustainability Impact Assessment of Land Use Changes, K. Helming; M. Pérez-Soba & P. Tabbush (Eds.), 405-424. Springer, Berlin.

Bradbury, R. (1996). Are indicators yesterday's news? Proceedings of the Fenner Conference "Tracking progress: Linking environment and economy through indicators and accounting systems", pp. 1-8. Sydney, University of New South Wales.

Brand, D.G. (1997). Criteria and indicators for the conservation and sustainable management of forests: progress to date and future directions. Biomass and Bioenergy, Vol. 13, Nos. 4/5, pp. 247-253.

Brockerhoff, E.; Jactel, H.; Parrotta, J.A.; Quine, C.P.; Sayer, J. (2008). Plantation forests and biodiversity: oxymoron or opportunity? Biodiversity and Conservation 17: 925-951.

Carnus, J-M.; Parrotta, J.; Brockerhoff, E.; Arbez, M.; Jactel, H.; Kremer, A.; Lamb, D.; O'Hara, K. & Walters, B. (2006). Planted forests and biodiversity. Journal of Forestry, Vol. 104, No. 2, pp. 65-77.

Castañeda, F. (2000). Criteria and indicators for sustainable forest management: international processes current status and the way ahead. Unasylva, Vol. 203 (51), No. 4, pp. 34-40.

Clark, M.R. & Kozar, J.S. (2011). Comparing sustainable forest management certifications standards: a meta-analysis. Ecology and Society, Vol. 16, No 1, Art. 3.

Cubbage, F.; Diaz, D.; Yapura, P. & Dube, F. (2010). Impacts of forest management certification in Argentina and Chile. Forest Policy and Economics, Vol. 12, pp. 497-504.

Dale, V.H. & Beyeler, S.C. (2001). Challenges in the develpment and use of ecological indicators. Ecological indicators, Vol. 1, pp. 3-10.

Elliot, C. (2000). Forest certification: A policy perspective. CIFOR, ISBN 979-8764-56-0. Bogor, Indonesia.

FAO (1995). Report of the expert meeting on the harmonization of criteria and indicators for sustainable forest management. Food and Agricultural Organization of the United Nations, Rome, Italy.

FAO (2010). Global forest resources assessment, 2010 – Main report. Food and Agricultural Organization of the United Nations Forestry Paper 163, Rome, Italy. Availability from
http://www.fao.org/forestry/fra/fra2010/en/ [Accessed November 2011]

FAO (2011). State of the World's forests 2011. Food and Agricultural Organization of the United Nations, Rome, Italy. Availability from
http://www.fao.org/docrep/013/i2000e/i2000e00.htm
[Accessed November 2011].

Forest Europe (2011) State of Europe's Forests 2011. Status and Trends in Sustainable Forest Management in Europe. Forest Europe, United Nations Economic Commission for Europe. Food and Agriculture Organization, Oslo, Norway, 337 pp. Availability from
http://www.foresteurope.org/pBl7xY4UEJFW9S_TdLVYDCFspY39Ec720-U9or6XP.ips
[Accessed February 2012]

FSC (2011). Global Forest Stewardship Council certifies: type and distribution. Availability from http://www.fsc.org/facts-figures.html [Accessed November 2011].

Gafo, M.; Caparros, A. & San-Miguel, A. (2011). 15 years of forest certification in the European Union. Are we doing things right? Forest Systems, Vol. 20, No. 1, pp. 81-94.

Gough, A.D.; Innes, J.L. & Allen, S.D. (2008). Development of common indicators of sustainable forest management. Ecological indicators, Vol. 8, pp. 425-430.

Hartley, M.J. (2002). Rationale and methods for conserving biodiversity in plantation forests. Forest Ecology and Management, Vol. 155, pp. 81-95.

Heikkinen, R.K.; Luoto, M.; Virkkala, R. & Rainio, K. (2004). Effects of habitat cover, landscape structure and spatial variables on the abundance of birds in an agricultural-forest mosaic. Journal of Applied Ecology, Vol. 41, pp. 824-835.

Hickey, G.M. & Innes, J.L. (2008). Indicators for demonstrating sustainable forest management in British Columbia, Canada: An international review. Ecological indicators, Vol. 8, pp. 131-140.

ITTO (1992). Criteria for the measurement of sustainable tropical forest management. International Tropical Timber Organization Policy Development. Series No. 3. Yokohama, Japan.

ITTO (1998). Criteria and indicators for sustainable management of natural tropical forests. International Tropical Timber Organization Policy Development. Series No. 7. Yokohama, Japan.

ITTO (2005). Revised ITTO criteria and indicators for the sustainable management of tropical forests including reporting format. International Tropical Timber Organization Policy Development. Series No. 15. Yokohama, Japan.

ITS Global (2011). Forest certification – Sustainability, governance and risk. International Trade Strategies, January 2011. Availability from http://www.itsglobal.net/sites/default/files/itsglobal/Forestry%20Certification-Sustainability%20Governance%20and%20Risk%20%282011%29.pdf   [Accessed November 2011].

Köhl, M.; Traub, G. & Päivinen, R. (2000). Harmonisation and standardisation in multi-national environmental statistics- mission impossible?. Environmental Monitoring and Assessment, Vol. 63, pp. 361-380.

Martres, J-L.; Carnus, J-M. & Orazio, C. (2011). Are MCPFE indicators suitable for planted forests? European Forest Institute Discussion paper No. 16. Availability from http://www.efi.int/files/attachments/publications/efi_dp16.pdf [Accessed November 2011]

McDonald, C.T. & Lane, M.B. (2004). Converging global indicators for sustainable forest management. Forest Policy and Economics. Vol. 6, pp. 63-70.

MCPFE (1993). General declaration and resolutions adopted. In: Proceedings of the Second Ministerial Conference on the Protection of Forest in Europe, Helsinki, 1993. Report. Liaison Unit, Vienna.

MCPFE (1998). General declaration and resolutions adopted. In: Proceedings of the Third Ministerial Conference on the Protection of Forests in Europe, Lisbon, 1998. Report. Liaison Unit Vienna.

MCPFE (2003). General declaration and resolutions adopted. In: Proceedings of the Fourth Ministerial Conference on the Protection of Forest in Europe, Vienna, 2003. Report. Liaison Unit, Vienna.

Michalos, A.C. (1997). Combining social, economic and environmental indicators to measure sustainable human well-being. Social Indicators Research, Vol. 40, pp. 221-258.

PECF (2011). Programme for the Endorsement of Forest Certification. Caring for our forests globally. Availability from http://pefc.org/about-pefc/who-we-are/facts-a-figures [Accessed November 2011].

Peng, C. (2000). Understanding the role of forest simulation models in sustainable forest management. Environmental Impact Assessment Review, Vol 20, pp. 481-501.

Poore, D. (2003). Changing Landscapes. Earthscan, ISBN 1-85383-991-4. London, UK.

Prabhu, R.; Ruitenbeek, H.J.; Boyle, T.J.B. & Colfer, C.J.P. (2001). Between voodoo science and adaptive management: the role and research needs for indicators of sustainable forest management. In: *Criteria and Indicators of Sustainable Forest Management*, R.J. Raison; A.G. Brown & D.W. Flinn (Eds.), 39-66. IUFRO research series, Vol. 7. CABI Publishing, Oxford.

Rametstenier, E. (2006). Opportunities to Create Synergy Among the C&I Processes Specific to the Topic of Harmonization. Inter-C&I Process Harmonization Workshop. Collaboration Among C&I Process – ITTO/FAO/MCPFE. Appendix 3 - Workshop Papers. Pp. 11-22. Bialowieza, Poland.

Rothertham, T. (2011). Forest management certification around the world – Progress and problems. The Forestry Chronicle, Vol. 87, No. 5, pp. 603-611.

Santiago Agreement (1995). Criteria and indicators for the conservation and sustainable management of temperate and boreal forests (The Montreal Process). Journal of Forestry. Vol. 93, No. 4, pp. 18-21.

Sprang, P. (2001). Aspects of quality assurance under the certification schemes FSC and PEFC. PhD thesis University of Freiburg, German. 70 pp. Availability from http://www.rainforest-alliance.org/forestry/documents/aspects.pdf [Accessed November 2011].

TAC (Technical Advisory Committee) (2007). Montreal Process Criteria and Indicators for the Conservation and Sustainable Management of Temperate and Boreal Forests. Technical notes on implementation of the Montreal Process Criteria and Indicators. Criteria 1-6. 2nd Edition. Availability from http://www.rinya.maff.go.jp/mpci/meetings/an-4.pdf [Accessed November 2011].

TAC (Technical Advisory Committee) (2009). Montreal Process Criteria and Indicators for the Conservation and Sustainable Management of Temperate and Boreal Forests. Technical notes on implementation of the Montreal Process Criteria and Indicators. Criteria 1-7. 3rd Edition. Availability from http://www.rinya.maff.go.jp/mpci/2009p_2.pdf [Accessed November 2011].

Teisl, M.F.; Peavey, S.; Newman, F.; Buono, J. & Hermann, M. (2002). Consumer reactions to environmental labels for forest products: A preliminary look. Forest Products Journal, Vol. 52, No. 1, pp. 44-50

Tomé, M. & Farrell, T. (2009). Special issue on selected results of the FORSEE project. Annals of Forest Science, Vol. 66, pp. 300.

UNECE/FAO (2011). Forest products. Annual market review 2010-2011. Geneve timber and forest study paper 27. United Nations, New York and Geneve. Availability from http://www.unece.org/fileadmin/DAM/publications/timber/FPAMR_2010-2011_HQ.pdf [Accessed November 2011].

USDA (2009). Conference Proceedings: Forest Criteria and Indicators Analytical Framework and Report Workshop. May 19-21, 2008 Joensuu, Finland. Gen. Tech. Report GTR-WO-81. USDA Forest Service, Washington Office. 350 p. Availability from http://treesearch.fs.fed.us/pubs/gtr/gtr_wo81.pdf [Accessed November 2011].

Van Dam, C. (2001). The economics of forest certification sustainable development for whom? Paper presented at The Latin American Congress on Development and Environment "Local Challenges of Globalization". Quito, Ecuador, April 11-12, 2003. Availability from http://cdc.giz.de/de/dokumente/en-d74-economics-of-forest-certification.pdf [Accessed November 2011].

Wang, S. & Wilson, B. (2007). Pluralism in the economics of sustainable forest management. Forest Policy and Economics, Vol. 9, pp. 743-750.

Wiersum, K.F. (1995). 200 years of sustainability in forestry: lessons from history. Environmental Management, Vol. 19, No 4, pp. 321-329.

Wijewardana, D. (2008). Criteria and indicators for sustainable forest management: The road traveled and the way ahead. Ecological indicators, Vol. 8, pp. 115-122.

WRM (2010). Tree monocultures in the South. World Rainforest Movement Bulletin Issue No 158, pp. 13-29. Availability from http://www.wrm.org.uy/index.html [Accessed November 2011].

Section 2

Carbon and Forest Resources

2

The Quality of Detailed Land Cover Maps in Highly Bio-Diverse Areas: Lessons Learned from the Mexican Experience

Stéphane Couturier
Laboratorio de Análisis Geo-Espacial, Instituto de Geografía, UNAM
Mexico

1. Introduction

The production of Land Use and Land Cover (LULC) maps is essential to the understanding of landscape dynamics in space and time. LULC maps are a tool for the measurement of human footprint and social processes in the landscape and for the sustainable use of finite resources on the planet, a growing challenge in our densely populated societies. LULC maps with detailed forest taxonomy constitute a basis for sustainable forest management, especially in highly biodiverse areas.

However, these maps are affected by misclassification errors, partly due to the intrinsic limitations of the satellite imagery used for map production. Misclassification occurs especially when categories of the classification system (classes) are not well distinguished, or ambiguous, in the satellite imagery. Therefore, statistical information on the quality, or accuracy, of these maps is critical because it provides error margins for the derived trends of land cover change, biodiversity loss and deforestation, these parameters being some of the few means that governmental agencies can provide as a guarantee of sustainable forest management practices associated with international conservation agreements.

Assessing the accuracy of LULC maps is a common procedure in geo-science disciplines, as a means, for example, of validating automatic classification methods on a satellite image. For regional scale LULC maps, because of budget constraints and the distribution of many classes over the large extension of the map, the complexity of accuracy assessments is considerably increased. Only relatively recently have comprehensive accuracy assessments, with estimates for each class, been built and applied to regional or continental, detailed LULC maps. However, the quasi totality of the cartography that has been assessed is for countries located in mainly temperate climates with low biodiversity. Instead, LULC maps in highly bio-diverse areas still lack this information, partly because their assessment faces uncertainty due to a high taxonomic diversity and unclear borders between forest classes.

This research focuses on the evaluation of the accuracy of detailed LULC regional maps in highly bio-diverse regions. These are provided by agencies of countries located in the sub-tropical belt, where no such comprehensive assessment has been done at high taxonomic resolution. This cartography is characterized by a greater taxonomic diversity (number of classes) than the cartography in low biodiversity areas. For example, in the United States of Mexico (USM, thereafter 'Mexico'), located in a 'mega-diverse' area, the map of the National

Forest Inventory (NFI) contains 75 LULC classes, including 29 forest cover classes, at the sub-community level of the classification scheme. Taxonomically, the NFI sub-community level in the USM is comparable to the subclass level of the National Vegetation Classification System (NVCS) of the USA, which contains 21 LULC classes, including 3 forest classes.

Higher taxonomic diversity, combined with highly dynamic landscapes, has several implications on the evaluation of errors. First, the numerous sparsely distributed classes represented in the classification scheme pose additional difficulties to the accuracy assessment of the map in terms of representative sampling. Second, thematic conceptual issues impact the way maps should be assessed, because more diversity introduces more physiognomic similarity among taxonomically close classes. As a result, more uncertainty is introduced in each label of the map as well as in each line of the map.

Confronted with these difficulties, this research presents a recently developed accuracy assessment framework, adapted to maps of environments with high biodiversity and highly dynamic landscapes. This framework comprises two methods derived from recent theoretical advances made by the geo-science community, and has been applied recently to the assessment of detailed LULC maps in four distinct eco-geographical zones in Mexico. The first method is a sampling design that efficiently controls the spatial distribution of samples for all classes, including sparsely distributed classes. The second method consists in a fuzzy sets-based design capable of describing uncertainties due to complex landscapes.

This chapter first describes the status of the accuracy assessment of LULC maps, an emerging branch of research in Geographical Information Science. Another section is focused on the methods employed for accuracy assessments of LULC maps and on the challenges related to the taxonomic diversity contained in maps of highly biodiverse areas. The next section focuses on the case of the Mexican detailed LULC cartography, as well as the framework that has been developed recently. Special emphasis lies on the distinctive features which make this case a pioneer experience for taxonomically detailed map assessments as well as a possibly valuable benchmark for other cartographic agencies dealing with biodiversity mapping in other regions of the world. Finally, the accuracy indices found for detailed LULC cartography in Mexico are presented and compared with the accuracy of other assessed international cartography. A major objective of this chapter is to appeal for the inclusion of accuracy assessment practices in the production of cartography for highly bio-diverse areas, because this kind of practice is still nearly absent to date.

2. Quality, or *accuracy* of land cover maps

2.1 Why is it important to measure the quality, or *accuracy* of a map?

A series of important applications typical of the sustainable management of land cover in bio-diverse areas relies on the information content of detailed Land Use/ Land Cover (LULC) maps: forest degradation and regeneration, biodiversity conservation, environmental services, carbon budget studies, etc. In many or all of these applications, map reliability and quality are usually unquestioned, given for granted, just as if each spatial unit on the map perfectly matched the key on the map, which in turn perfectly matched reality. The minimum mapping unit, which defines the scale of the map, is commonly the only information available about the spatial accuracy of a map and no statistically grounded reliability study is applied as a plain step of the cartographic production process.

In general, the comprehensive LULC cartography of a region is obtained through governmental agencies of a country or group of countries, at regional scale, intermediate between local (> 1:50,000) and continental (1:5,000,000). Since the 1990s, the classification of satellite imagery has become the standard for LULC mapping programs at regional scale. However, the classification process is affected by different types of error (Couturier et al., 2009a; Green & Hartley, 2000) related in part to the limited discrimination capacity of the spaceborne remote sensor. The difficult distinction, on the satellite imagery, between categories (or 'thematic classes') of a cartographic legend can cause a high percentage of errors on the map (see next subsection), especially on maps with high taxonomical detail (high number of thematic classes). This is why a forest management policy or a biodiversity monitoring program whose strategy is simply 'process map information and rely on the quality of the map' is highly questionable.

For example in highly bio-diverse regions within Mexico, typical comprehensive database and cartographic products, such as the cartography generated by the National Institute of Statistics, Geography and Informatics (INEGI) and CONAFOR (the National Commission for Forests), are obtained at scale 1:250,000. However all of these products remain deprived of statistical reliability study. This is most unfortunate since the latter governmental agency produces statements on recent deforestation rates based on these maps (online geoportal: CONAFOR, 2008), and these statements, because of the absence of statistical reliability study, remain the focus of distrust and controversial academic and public discussions. It is worth stating that the online availability of the satellite imagery – a feature advertized by this governmental agency - does *not* make an index derived from the imagery more reliable. The extraction of the index based on colour tones of the satellite imagery available online is far from trivial and it is simply impossible for a user to quantitatively derive the global reliability of the cartography out of internet access to the imagery.

An error bar is sometimes present aside the legend of INEGI maps and indicates an estimate of positional errors in the process of map production. However, the procedure leading to this estimate is usually undisclosed, and any objective interpretation of this estimate by the user is thus discouraged (Foody, 2002). Moreover, such error bar indicates a very reduced piece of information with respect to the thematic accuracy of the map.

Instead, the *accuracy* of a cartographic product is a statistically grounded quantity which gives the user a robust estimate of the agreement of the cartography with respect to reality. Such estimate is essential when indices derived from cartography – i.e. spatial extent statistics, deforestation rates, land use change analysis - are released to the public or to intergovernmental environmental panels, while the absence of such estimate indicates that these indices stand without error margins, and as such, without statistical validity. The accuracy of a map also serves as a measurement of the risk undertaken by a decision maker using the map. Besides, this information allows error propagation modeling through a Geographical Information Systems or GIS (Burrough, 1994) in a multi-date forest monitoring task, for example. The construction of the accuracy estimate is generally named 'accuracy assessment' and is explained in section 3.

2.2 Status of the measured accuracy of land cover maps

Assessing the accuracy of LULC maps is a common procedure in geo-science disciplines, as a means, for example, of validating automatic classification methods on a satellite image. For regional scale LULC maps, because of budget constraints and the distribution of many classes over the large extension of the map, the complexity of accuracy assessments is considerably increased. Only relatively recently have comprehensive accuracy assessments, with estimates for each class, been built and applied to regional or continental, detailed

LULC maps. In Europe, Büttner & Maucha (2006) reported the accuracy assessment of 44 mapped classes (including 3 forest classes) of the CORINE Land Cover (CLC) 2000 project. In the United States of America (USA), Laba et al. (2002) assessed the accuracy of 29 LULC classes and Wickham et al. (2004) the accuracy of 21 classes in maps of year 1992 from, respectively, the Gap Analysis Project (GAP) and the National Land Cover Data (NLCD). As a part of the Earth Observation for Sustainable Development (EOSD) program of Canada, Wulder et al. (2006) provide a plan for the future accuracy assessment of the 21 classes in the 2000 Canadian forest cover map, and the accuracy of this program is assessed in the Vancouver Island for 18 classes (Wulder et al., 2007).

These studies reveal the presence of numerous confusions between classes, which yield a global accuracy index (percent area of the map with correct information) of between 38 and 70%. Consequently, these reliability studies constitute very valuable information in terms of the practical use of the assessed maps as well as in terms of enhanced map production strategies in the future.

The cartography of countries situated in areas of high bio-diversity is characterized by a greater taxonomic diversity, i.e. a greater number of classes for a given taxonomic level, than the above cited cartography. However, as is currently the case of the quasi totality of the countries situated in areas of high bio-diversity, the Mexican NFI map, for example, was until recently deprived of statistically grounded information on its reliability. Table 1 reports a collection of 9 studies in the world where a statistically grounded accuracy assessment has been applied to regional LULC cartography. This collection is thought to be relatively representative of existing studies and therefore reflects the status of international accuracy assessments of regional LULC maps to date. The studies which employ a probabilistic sampling design in the sense of Stehman (2001) over the entire area and not just a partial sampling are highlighted in bold. The list of studies was sorted according to the thematic richness of the assessed map (total number of classes).

Some findings can be derived from this table; for example, at first sight, the assessment efforts seem to be greater on the American continent than in other places. The LULC cartography on the African continent is represented by a study with partial assessment in Nigeria; the regional cartography derived from the Africover 2000 project (part of the Global Land Cover, or GLC, project) has not yet been submitted to a probabilistic accuracy assessment to date. In terms of taxonomic diversity (number of mapped classes), the 2000 NFI map in Mexico ranks second after the Southwest USA map, and ranks first of the mega-diverse areas. Therefore, the study on the 2000 NFI map in Mexico stands out as especially important in the world. Among the probabilistic assessments, the study assesses the highest number of classes (32 assessed classes vs 22 in Europe which holds the second ranking). For comparison purposes, we indicated the equivalent taxonomic level of each map, with respect to the four aggregation levels (*biome, type, community, community with alteration*, also known as *sub-community*) considered for the classification system of the IFN 2000 (Palacio-Prieto et al., 2000), plus two more detailed levels (*community with density grades* and *association with alteration*). The taxonomic level of the maps is generally relevant to applications of regional forest management and biodiversity monitoring (7 studies involve maps of levels *community, community with alteration, community with density grades, association with alteration*, which are the most detailed taxonomic levels). However, the study on the NFI 2000 map is the only accuracy assessment *per class* of these levels of taxonomic detail in a mega-diverse area (the other detailed assessments are in the USA, Europe and Canada), a level of detail which actually allows statistically- based cartographic management schemes in terms of bio-diversity dynamics. Another noteworthy study in a mega-diverse area is the one in South and Southeast Asia (Stibig et al., 2007), but its accuracy assessment was only obtained at the biome level. A study at the biome level does allow a deforestation study (forest – non forest change) with error margins, but does not allow a land cover change study with more detailed processes (e.g. 'forest to forest with alteration'), also important in sustainable management.

However, the assessment of the NFI 2000 cartography in four eco-geographical areas only constitutes a pilot study, confined to a limited extension, in a mega-diverse area. The spatial extent subject to the assessment is about 19,500 km², much smaller than the majority of the other studies (seven of the nine studies involve extents of more than one million km²). Indeed, the enhanced taxonomic diversity, combined with highly dynamic landscapes, increase the difficulty of the accuracy assessment of maps in mega-diverse areas (Couturier et al., 2007), a fact that probably contributes to explain the lack of studies in such areas.

3. How can I measure the quality, or *accuracy* of land cover maps in biodiverse areas?

Generally, map accuracy is measured by means of reference sites and a classification process more reliable than the one used to generate the map itself. The classified reference sites are then confronted with the map, assuming that the reference site is "the truth". Agreement or disagreement is recorded in error matrices, or confusion matrices (Card, 1982), on the base of which various reliability indices may be derived. For regional scale LULC maps, the abundance and distribution of classes over the large extension of the map, confronted with tight budget constraints, add complexity to accuracy assessments. Only relatively recently have comprehensive accuracy assessments, with estimates for each class, been built and applied to regional or continental LULC maps (e.g. Laba et al., 2002; Stehman et al., 2003; Wickham et al., 2004; Wulder et al., 2007). Because of the high complexity of these products, detailed information on the assessment process itself is needed for the reliability figures to be interpreted properly (Foody, 2002). With this understanding, Stehman & Czaplewski (1998) have proposed a standard structure for accuracy assessment designs, divided into three phases:
1.   Representative selection of reference sites (sampling design),
2.   Definition, processing and classification of the selected reference sites (verification design),
3.   Comparison of the map label with the reference label (synthesis of the evaluation).
Wulder et al. (2006) provide a review on issues related to these three steps of an accuracy assessment design for regional scale LULC cartography. We indicate in the next sub-section the features and techniques most commonly employed in the literature for phases 1-3.

3.1 Methods employed in the accuracy assessment of LULC maps in the world
3.1.1 Sampling design
The first phase of the accuracy assessment is the sampling design. The selection of the reference sites is a statistical sampling issue (Cochran, 1977), where strategies have varied according to the application and complexity of the spatial distribution. Stehman (2001) defines the probability sampling, where each piece of mapped surface is guaranteed a non-null probability of inclusion in the sample, as being a basic condition for statistical validity. In most local scale applications, reference sites are selected through simple random

Region of the world	Acronym of project and year of cartography	Prevailing biotic environment	Assessment design	Number of classes Total	Number of classes Forest	Equivalent taxonomic level	Spatial resolution and satellite sensor	Spatial extent of the map effectively assessed	Reference publication
Southwest USA	GAP 2000	Temperate-dry	Partial (near to roads)	125 (85 assessed)	27 (18 assessed)	Association with alteration	30m (Landsat TM)	? (1.4 M km²)	Lowry et al. (2007)
Mexico (4 areas)	NFI 2000	Mega-diverse	Probabilistic	75 (32 assessed)	29 (19 assessed)	Community with alteration	1km (Landsat TM)	0.0195 M km² (1.95 M km²)	Couturier et al. (2010)
European Union	CorineLC 2000	Temperate	Probabilistic	44 (22 assessed)	3	Community	30m (Landsat TM)	2.68 M km²	Buttner and Maucha (2006)
South and Southeast Asia	TREES 2000	Mega-diverse	Probabilistic for biome level	40 (= 4 biome classes assessed)	17 (only 'forest' class assessed)	Community with alteration	1km (SPOT-VEGETATION)	4.5 M km²	Stibig et al. (2007)
India	ISRO-GBP 1999	Mega-diverse	Partial (in 3 states of the country)	35	14	Community	188m (WiFS, IRS)	? (3.3 M km²)	Joshi et al. (2006)
USA	NLCD 1992	Temperate	Probabilistic	21	3	Community	30m (Landsat TM)	9.1 M km²	Stehman et al. (2003)
Canada (1 area)	EOSD- Forest 2000	Temperate	Probabilistic	18	10	Community with density	25m (Landsat TM)	0.031 km²	Wulder et al., (2007)
Nigeria	1990	Tropical humid	Partial	8	3	Biome	1km (NOAA AVHRR)	? (0.904 km²)	Rogers et al. (1997)
Legal Amazon, Brasil	GLC 2000	Tropical humid and dry	Partial	5	3	Type	1km (SPOT-VEGETATION)	? (5.0 M km²)	Carreiras et al. (2006)

USA: United States of America.

Prevailing biotic environment: If large areas of different environments exist, e.g. temperate and sub-tropical, dry and/or humid, we indicated 'mega-diverse'.

Assessment design: 'probabilistic' if the design is associated with the total area mapped, and 'partial' if not; the design is probabilistic according to criteria of statistical rigor established by Stehman (2001).

Equivalent taxonomic level: equivalence in terms of the classification system of the National Forest Inventory 2000 in Mexico. Taxonomic levels are *Biome ('Formación' in Spanish)*, *Type*, *Community*, *Community with alteration* (or *sub-community*), sorted from the most general to the most detailed (Palacio-Prieto et al.: 2000). Two additional more detailed levels were considered: *Community with density* (vegetation density levels) and *Association with alteration*.

Spatial extent of the map effectively assessed: M km2 = millions of square kilometers. In case of partial assessment designs, many authors do not report sufficient information that indicate the effective area actually assessed; if this is the case we indicate '?', and the figure between parenthesis represents the total extent of the map (not the one actually assessed).

Table 1. List of major published studies on assessments of regional land use/ land cover maps in the world. The list is relatively exhaustive of institutional programs which aim a probabilistic sampling design. The studies are sorted according to the total number of classes contained in the legend of the map.

sampling. Two stage (or double) random sampling has been preferred in many studies in the case of regional cartography; in a first step, a set of clusters is selected through, for example, simple random sampling. This technique permits much more control over the spatial dispersion of the sample, which means much reduction of costs (Zhu et al., 2000), and was adopted for the first regional accuracy assessments in the USA, for LULC maps of 1992 (Laba et al., 2002; Stehman et al., 2003).

A random, stratified by class sampling strategy means that reference sites are sampled separately for each mapped class (Congalton, 1988). This strategy is useful if some classes are sparsely represented on the map and, therefore, difficult to sample with simple random sampling. This strategy was adopted at the second stage of their double sampling by Stehman et al. (2003) and Wickham et al. (2004).

Systematic sampling refers to the sampling of a partial portion of the mapped territory, where the portion has been designed as sufficiently representative of the total territory. This strategy, adopted as a first stratification step, is attractive for small scale datasets and reference material of difficult access: Wulder et al. (2006) define a systematic stratum for the future (and first) national scale accuracy assessment of the forest cover map in Canada.

3.1.2 Verification design

For regional scale detailed land cover maps, the frame for reference material of phase 2 is typically an aerial photographic coverage (e.g. Zhu et al., 2000), and ground survey is only occasional. For all studies cited in the text of section 3 so far, the classification of reference sites was based on more precise imagery i.e. imagery with higher spatial resolution, than the imagery that was employed during the map production process. In these cases the map was produced using Landsat imagery (spatial resolution of 30m), and was assessed using aerial photographs (spatial resolution better than 3m) or aerial videography (Wulder et al., 2007). The map of South and Southeast Asia (Stibig et al., 2007, table 1) was produced using the SPOT-VEGETATION sensor (spatial resolution of 1km) and assessed using Landsat imagery (resolution 30m). An alternative reference material for recent LULC cartography could be a wide coverage of very high resolution satellite imagery such as the one available on the online Google Earth database. For all studies, remote sensing based reference data has been preferred as the primary material instead of ground survey for its cost-effectiveness in large areas.

Double sampling techniques are effective at controlling the spatial dispersion of the sample among image/ photograph frames if these are taken as the cluster, or Primary Sampling Unit (PSU), for first stage sampling (see previous subsection).

Congalton & Green (1993) relate errors of the map to imprecise delineation and/or misclassification. Additionally, the imperfect process of the assessment itself also generates erroneous statements on whether the map represents reality or not. A main topic is the positional error of the aerial photograph with respect to the map. To this respect, a procedure ensuring geometric consistency must be included in the evaluation protocol. For example, the procedure of visually locating sample points on the original satellite imagery, described in Zhu et al. (2000), reduces the inclusion of errors due to geometric inconsistencies. Other sources of fictitious errors occur in phase 3 (labeling protocol), and are related to the thematic and positional uncertainties of maps. This topic is introduced in section 3.2 and fully devised in Couturier et al. (2009a).

3.1.3 Synthesis of the evaluation
The comparison between the information contained on the map and the information derived from the reference site yields an agreement or a disagreement. Typically, the numbers of agreements and disagreements are recorded and form a confusion matrix. However, these numbers are reported in the matrix with weights that depend on the probability of inclusion of the reference site in the sample (Stehman, 2001). This probability of inclusion is defined by the sampling design. For example, a simple random selection is associated with a uniform (constant) inclusion probability among all reference sites. For a two stage sampling, the probability of inclusion follows Bayes law: The probability of inclusion p_{2k} of a reference site at the second stage is a multiplicative function of the inclusion probability p_{1k} of the cluster it pertains to, and of the inclusion probability of the reference site, once the cluster has been selected $p_{2|1}$ (conditional inclusion probability)(equation 1):

$$p_{2k} = p_{2|1} * p_{1k} \qquad (1)$$

Accuracy indices per class are derived from these calculations: 'user's accuracy' of class k is the account of agreements from all sites of mapped class k while the 'producer's accuracy' of class k counts agreements from all reference sites labeled as class k. The respective disagreements correspond to 'commission errors' and 'omission errors' (Aronoff, 1982). The global accuracy index, or proportion correct index, which indicates the accuracy of the map as a whole (all thematic classes), integrates the accuracy level of all classes, weighted by the probability of inclusion specific to each class. In this calculation, weights usually correspond to the relative abundance of the class on the map. Other reliability indices are popular, such as the Kappa index, which takes into account the contribution of chance in the accuracy (Rosenfield and Fitzpatrick-Lins, 1986). However, in regional scale accuracy assessments, the proportion correct indices are preferred, because they are coherent with the interpretation of confusions according to area fractions of the map (Stehman, 2001).

A confidence interval of the accuracy indices can be estimated, although only few accuracy assessments provide this information. A popular estimate of the confidence interval is based on the binomial distribution theory: the confidence interval of the accuracy estimate depends on the sample size and on the reliability value (accuracy estimate) in the following manner (Snedecor & Cochran, 1967, cited by Fitzpatrick-Lins, 1981)(equation 2):

$$d^2 = t^2 p (1-p) / n \qquad (2)$$

where d is the standard deviation (or half the confidence interval) of the estimate, t is the standard deviate on the Gaussian curve (for example, t = 1.96 for a two-sided probability of error of 0.05), p is the reliability value, and n is the number of sampled points. Although most accuracy assessments refer to it, this binomial distribution formula is only valid for simple random sampling. For more sophisticated sampling designs (e.g. two stage sampling) the confidence interval is influenced by the variance of agreements among clusters. Estimators integrating inter-cluster variance (Stehman et al., 2003) are seldom employed in map accuracy assessments because of their complexity (Stehman et al., 2003). For the cartography assessment in Mexico, an estimator which includes an inter-cluster variance term was used in Couturier et al. (2009a). The estimate was built on a stratified by class selection in the second-stage of the sampling design (Särndal et al., 1992).

3.2 Methodological challenge for the accuracy assessment of detailed LULC maps

The detailed cartography of highly bio-diverse regions is characterized by a greater taxonomic diversity (number of classes) than the cartography of regions in mainly temperate climates. Greater taxonomic diversity, combined with highly dynamic landscapes, has several implications on the evaluation of errors.

First, the numerous sparsely distributed classes represented in the classification scheme pose additional difficulties to the accuracy assessment of the map in terms of representative sampling.

Second, thematic conceptual issues impact the way maps should be assessed, for reasons illustrated in three cases:

- More diversity introduces more physiognomic similarity among taxonomically close classes: for example, cedar forest is an additional conifer forest class in sub-tropical environments, so mixed conifer forest patches are more difficult to classify, and boundaries between conifer forests are more difficult to set. As a result, more uncertainty is introduced in each label of the map as well as in each line of the map.

- Highly dynamic landscapes mean more classes placed along a continuum of vegetation, where some classes are a temporal transition to other classes. For example, the sequence of classes 'pasture to secondary forest to primary forest' is characteristic of sub-tropical landscapes. The extremes of such sequence may be spectrally distinct and easily separated, however boundaries between intermediate classes are difficult to interpret.

- More diversity combined with highly dynamic landscapes means more fragmented landscapes composed of small patches of different classes. The interpretation of these results in heterogeneous patches is difficult to assess.

Third and last, ambiguity between classes on satellite imagery, related to the above situations, becomes more likely. In these conditions, the information on spectral separability could be a systematic tool to prioritise future cartographic efforts (Couturier et al., 2009b).

Confronted with the three implications, we developed two methods based on recent theoretical advances made by the geo-science community.

The first method comprised a sampling design that efficiently controlled the spatial distribution of samples for all classes, including sparsely distributed (or fragmented) classes. Previous assessments have relied on two-stage sampling schemes where simple random or stratified by class random sampling was employed in the first stage. Couturier et al. (2007) demonstrated that these strategies fail in the context of the Mexican NFI. Section 3.2.1 presents a two-stage hybrid scheme where proportional stratified sampling is employed for sparsely distributed (rare) classes. This scheme was applied to four areas in distinct eco-geographical zones of Mexico (see section 4.3).

The second method was to design a fuzzy sets-based design capable of describing uncertainties due to complex landscapes. We will see in section 3.2.2 that it is traditionally possible to incorporate a thematic fuzzy component in accuracy assessment designs, but this component, as well as positional uncertainty, are implicitly fixed by the map producer, with no possible change after the design has been applied. Recently, advances in fuzzy classification theory have permitted the comparison of maps incorporating thematic and positional uncertainties.

3.2.1 The sampling design for fragmented (rare) classes

In order to find a sampling design well suited to an abundant set of fragmented, sparsely distributed (or rare) classes, several double sampling designs (DS) were previously tested in a pilot study, the closed watershed of the Cuitzeo lake in Mexico (Couturier et al., 2007);

DS1 was defined as the simple random selection of the Primary Sampling Units (PSUs), as in Laba et al. (2002).

DS2 was characterized by the random, stratified by class, selection of PSUs, as in Stehman et al. (2003).

DS3 was defined as a proportional, stratified by class, selection of PSUs. For the latter design, not applied in previously published research, the probability of inclusion of a PSU is proportional to the abundance of the class in the PSU. The abundance of a class equates its area fraction, easily obtainable via attribute computation in a GIS. Then, the probability of inclusion of Secondary Sampling Units (SSUs) at the second stage was defined as being inversely proportional to the abundance of the class in the PSU. Proportional sampling is a known statistical technique (Cochran, 1977) and some characteristics of its application to map accuracy assessment are devised in Stehman et al. (2000). However, DS3 had never been applied in published studies, maybe because it was not necessary for maps with classification systems of mainly temperate countries.

Finally, an entirely novel, hybrid design (DS4) includes a simple random selection of PSUs (as in DS1) for common classes (area fraction above 5%, 7 classes in Cuitzeo), and a proportional stratified selection of PSUs (as in DS3) for rare classes (area fraction below 5%, 14 classes in Cuitzeo). After selection of the PSUs, the sample size of SSUs was fixed at 100 per mapped class, a value widely adopted in similar assessments (Stehman & Czaplewski, 1998).

With fixed operational costs, the only design that systematically provided statistically representative estimates for all classes was the hybrid design DS4 (Couturier et al., 2007). Additionally, the hybrid design achieved a spatial dispersion of the sample similar to the dispersion achieved by DS1, with simple random selection of Primary Sampling Units (PSUs). DS1 is known for generating a good dispersion of the sample in regional map assessments. For this reason, DS1 was successfully applied in the accuracy assessment of the NLCD project in the USA (Stehman et al., 2003). However, DS1 was discarded in our pilot study because it was not able to handle the high number of rare classes of the NFI. Instead, the hybrid design maintains simple random selection of PSUs for common classes, but applies a proportional stratified selection of PSUs for rare classes. This way, DS4 cumulates the advantages of a wide-spread sample dispersion for common classes, and the advantages of a sufficient sample size and easy estimate calculation for rare classes.

3.2.2 The fuzzy approach for positional and thematic uncertainties

In traditional accuracy assessment, the labeling protocol (phase 3 of the accuracy assessment) consists in attributing one and only one category of the classification scheme to each reference site. However, this procedure assumes that each area in the map can be unambiguously assigned to a single category of the classification scheme (or LULC class). In reality, the mapped area may be related to more than one LULC class because of the characteristics of the landscape in the reference site. This conceptual difficulty is ignored in the traditional (or Boolean) labeling protocol, and may conduce to an under-estimation of map accuracy (Foody, 2002). In particular, this difficulty arises in the following cases:

- The landscape in the reference site has physiognomic similarities with more than one LULC class. For example, a one hectare forest patch containing oak trees and two or three pine trees has physiognomic similarity with forest class 'oak forest' and forest class 'oak-pine forest'. As a result, the map label for this site is affected with uncertainty. The reference site could be in a transition zone between an oak forest and a oak-pine forest.

• The landscape in the reference site is a patch within a continuum of vegetation, where the LULC classes represented are a temporal transition to other classes. For example, the sequence of classes 'pasture to secondary forest to primary forest' is characteristic of some sub-tropical landscapes. As a result, the map label for this site is affected with uncertainty. The extremes of such sequence may be easily identified on the ground, however boundaries between patches of intermediate classes are difficult to set. As a result, lines between mapped objects for this site are affected with uncertainty.

• The landscape in the reference site is a fragmented landscape, composed of small patches (below minimum mapping unit) of different land use or land cover. The interpretation of this mixed reference site, because of the scale of the map, must be a non unique label. As a result, the map label for this site is affected with uncertainty.

Due to the above described continuous or fragmented aspects of land use and land cover in a landscape, maps with discrete representation (discrete, or crisp, class assignation) and infinitely small line features (crisp boundaries of objects) necessarily describe reality with a certain margin of uncertainty. In order to take this uncertainty aspect into account, it has been referred to the concept of fuzzy sets (Zadeh, 1965).

In the crisp approach, an element x of the map X belongs totally to a class k of the set C or does not belong to it. A way of representing this is to define a membership function μ, which takes the value '1' if the element x belongs to class k and '0' otherwise. This assignation process can be called Boolean labeling. In a typical case of photo-interpretation for map accuracy assessment, a forest reference site with a crown cover close to 40% may pertain to a transition zone between closed forest (crown cover > 40%) and open forest (crown cover < 40%). If the photo-interpreter characterizes this site as closed forest and the corresponding label on the map is open forest, then this site is interpreted as an error on the map.

In fuzzy sets theory, an element belongs to a set or class with a certain degree of similarity, probability or property, some of these notions being contained in a 'degree of membership', depending on the application. One element x may belong to various classes at a time with different degrees of membership $\mu_k(x)$. For example, quantitative degrees of membership take a value between 0 and 1 to express the partial membership to various classes of the set. With this approach, the reference site with a tree cover close to 40%, would be characterized for instance by a 0.5 degree of membership in both classes (open and closed forest).

Many authors have rejected the term "fuzzy set theory" to characterize landscape interpretation, in favor of "soft" or "continuous" classification. Critiques have noted that the use of a continuous range of membership values does not entail employment of the concepts of fuzzy logic (Haack, 1996). Nevertheless, the term "fuzzy classification" will be used here as a compromise, recognizing the heritage of these techniques but emphasizing the classification process over the logic of set theory.

Cartographical models that present a fuzzy classification approach were developed (Equihua, 1990, 1991; Fisher & Pathirana, 1990; Foody, 1992; Wang, 1990). These models allow the representation of the landscape features previously enumerated in this subsection. Despite the perspective of a more lawful representation of real landscapes, these models present two limitations:

• The interpretation and manipulation of fuzzy classified maps by users already accustomed to crisp maps is still a pending challenge; each point on the map represents various LULC classes with different degrees of membership. The vast majority of maps, including the existing LULC maps in Mexico and in territories with high biodiversity, are crisp.

- The coherent production of fuzzy classified maps with quantitative degrees of membership is not possible in all mapping situations. One of the situations where such fuzzy classified map can be easily produced is a binary map of, for example, forest/non-forest where percent crown coverage represents one of the fuzzy labels. A second situation is a map made of ordinal categories, where uncertainty between categories can be modeled by a fuzzy matrix (illustrated in Hagen, 2003). A third situation occurs when automatic processing is constructed so as to generate the quantitative fuzzy labels. A typical example of this third situation is a map of unmixed fractions of LULC classes, extracted from automatic spectral mixing analysis, where the classes are represented by pure end-member pixels. However, the assignment of quantitative fuzzy labels during visual interpretation, for example, can be affected by subjectivity. This is possibly a reason why quantitative fuzzy labeling has generally not been adopted in mapping situations with visually interpreted material.

Consequently, for the challenge concerning land cover over highly bio-diverse regions, the focus was made on assessing a crisp map with fuzzy classified reference material. As mentioned in section 3.1, the typical reference material of regional accuracy assessments is aerial photographs. We were confronted with the subjectivity of interpreters in preliminary attempts at classifying the material with quantitative degrees of membership. For these reasons, we settled for the fuzzy classification technique expressed by linguistic rules, introduced for visual interpretation by Gopal & Woodcock (1994), and commonly employed. This technique of fuzzy classification is described in the verification design of Couturier et al. (2008) for the case of detailed land cover map assessment in Mexico.

The use of fuzzy classification techniques in the labeling protocol permits the reduction of fictitious errors in the process of map assessment, fictitious errors being due to the thematic uncertainty of maps. However, as said earlier, maps are also characterized by positional uncertainty. This uncertainty may also affect the accuracy results when the assessed map is compared with the reference material. As a result of advances in fuzzy classification theory, much research have focused on the comparison of fuzzy classified maps and on the multi-scale comparison of maps (Pontius & Cheuk, 2006; Remmel & Csillag, 2006; Visser & de Nijs, 2006). In Couturier et al. (2009a), the systematic inclusion of positional uncertainty within regional accuracy assessments is proposed, formalized, and applied to the case of land cover maps of highly bio-diverse regions.

4. Mexican detailed land cover cartography and the application of the methods developed recently

4.1 Mexican detailed LULC cartography

As a consequence of its extension over a wide range of physio-graphical, geological and climatic conditions, the Mexican territory is composed of a remarkably large variety of ecosystems and diversity of flora (Rzedowski, 1978), is among the five richest countries in biological diversity and therefore considered as a mega-diverse area (Velázquez et al., 2001). In turn, this range of environmental conditions predetermined transformations of the landscape by humans in a variety of ways. The intensification of land uses over the last century and the response of the eco-systems to this intensification altogether shaped the complex landscapes in the contemporary Mexico.

In the past three decades, governmental agencies in the North American sub-continent have promoted the production of geographic information at a regional scale, which we define intermediate between continental (1:5 000 000) and local (> 1:50 000). The major historical data set of regional scale (1:250 000) LULC maps in Mexico was developed by the National Institute of Statistics, Geography and Informatics (INEGI). In the nineteen eighties, the first set of 121 LULC maps was published for the entire territory, based on the interpretation of aerial photography collected from 1968 to 1986 (average date 1976) and considerable ground work (INEGI, 1980). This dataset was part of the INEGI first series ('INEGI serie I', in Spanish) cartography. In the mid nineteen nineties, INEGI produced the second series cartography ('INEGI serie II') in a digital and printed format. The LULC maps of INEGI serie II were elaborated using the former series I maps, and visual interpretation of Landsat Thematic Mapper (TM) images acquired in 1993, printed at scale 1:250 000. The INEGI cartography legend included 642 categories to consistently describe LULC in the entire country. For land cover categories, or classes, this detailed classification scheme was based on physiognomic, floristic and phenological attributes of plant communities (table 1) and degrees of anthropic modification.

Formation	Vegetation Types
Temperate Forest	1. Cedar forest , 2. Fir forest, 3. Pine forest, 4. Conifer scrubland, 5. Douglas fir forest, 6. Pine-oak woodland, 7. Pine-oak forest , 8. Oak-pine forest, 9. Oak forest, 10. Mountain cloud forest, 11. Gallery forest.
Tropical forest	*Perennial & sub-perennial tropical forests:* 12. Tropical evergreen forest, 13. Tropical sub-evergreen forest, 14. Tropical evergreen forest (medium height), 15. Tropical sub-evergreen forest (medium height), 16. Tropical evergreen forest (low height), 17. Tropical sub-evergreen forest (low height) , 18. Gallery forest.
	Deciduous & sub-deciduous forests: 19. Tropical sub-deciduous forest (medium height), 20. Tropical deciduous forest (medium height), 21. Tropical sub-deciduous forest (low height), 22. Tropical deciduous forest (low height), 23. Tropical spiny forest.
Scrubland	24. Sub-montane scrubland, 25. Spiny Tamaulipecan scrubland, 26. Cacti-dominated scrubland 27. Succulent-dominated scrubland, 28. Succulent-cacti-dominated scrubland, 29. Sub-tropical scrubland, 30. Chaparral, 31. Xerophytic scrubland, 32. Succulent-cactus-dominated cloud scrubland,, 33. Rosetophilous scrubland, 34. Desertic xerophytic rosetophilous scrubland, 35. Desertic xerophytic microphilous scrubland,, 36 Propospis spp.-dominated, 37. Acacia spp.-dominated, 38. Vegetation of sandy desertic conditions.
Grassland	39. Natural grassland, 40. Grassland-huizachal, 41. Halophilous grassland, 42. Savannah, 43. Alpine bunchgrassland, 44. Gypsophilous grassland.
Hygrophilous vegetation	45. Mangrove, 46. Popal-Tular (hygrophilous grassland), 47. Riparian vegetation.
Other vegetation Types	48. Coastal dune vegetation, 49. Halophilous vegetation.

Table 2. Classification scheme of the INEGI land use and vegetation cartography (only natural land cover categories are indicated):

In the year 2000, the Ministry of the Environment in Mexico (SEMARNAP) attributed the task of updating the LULC map of the country (at scale 1:250 000) to the Institute of Geography of the Universidad Nacional Autónoma de México (UNAM). This task was intended as an academic-driven methodological proposal for rapid nation-wide detailed forest assessments. In this perspective, the cartographic project was named the National Forest Inventory (NFI) map of year 2000. An important objective of the project was the compatibility with previous cartography in view of LULC change studies. Rapidity (8 months) and low cost of execution were constraints that guided the planning of the project.

Visual interpretation of satellite imagery, with the aid of INEGI previous LULC digital cartography, was selected as the best classification strategy. However, the classification scheme was adjusted to the capacity of the Landsat Enhanced TM plus (ETM+) imagery at discriminating classes, according to previous classification experience in Mexico (e.g. Mas & Ramírez, 1996). The 642 categories of the INEGI cartography legend (including 49 vegetation types in table 2) were aggregated into 75 thematic classes (community level, with the inclusion of two levels of human induced modification) and further into three coarser levels of aggregation.

Visual interpretation was done on ETM+ imagery of the drier season, acquired between November 1999 and April 2000. The best option for interpretation was visually selected among various colour composites. The methods and results of the IFN 2000 cartographic project have been published (Mas et al., 2002a; Palacio-Prieto et al., 2000; Velázquez et al., 2002). Figure 1, taken from Mas et al. (2002a), illustrates the 2000 NFI map at formation level (coarsest level of aggregation). The present research focuses on the cartographic product with the finest level of aggregation (community level, with the inclusion of degradation levels), because of the availability of abundant quasi synchronous aerial photograph cover all throughout the country which can be used as independent reference data for accuracy assessment.

Since 2001, the National Commission of Forests (CONAFOR), an agency dependent of the National Environmental Agency in Mexico (SEMARNAT), is in charge of updating the vegetation cover change in Mexico, in parallel with the INEGI regional LULC cartography (year 2002: 'Serie III' map, and year 2007: 'Serie IV' map). None of this cartography so far has been generated with an international standard accuracy assessment scheme as described in this chapter. Since 2004, CONAFOR has established a 5 year repeat forest inventory of the Mexican territory ('Inventario Nacional Forestal y de Suelos', INFyS, 2008), based on a systematic grid of ground plots over the entire vegetation cover of Mexico.

4.2 Developing the framework for assessing the Mexican detailed LULC cartography

As stated previously, if we except the material presented in this chapter, all detailed LULC cartography in Mexico is characterized by the absence of quantitative, reliable information on its quality. Consequently, only qualitative statements can characterize the reliability of archive and recent Mexican cartographic products, based on a judgment on the quality of the data that was employed in the map production process. For example, the INEGI serie I data (1968-1986) are expected to be very reliable in terms of thematic accuracy, because of the quality of the field reference data, but their temporal coherence (accuracy) is low. Conversely, the LULC maps of INEGI serie II are characterized by a high temporal coherence. However, because the visual interpretation of only one colour composite of

Landsat imagery (bands 4,3,2) was used to update a map with very high taxonomic precision (INEGI legend of 642 classes), the thematic accuracy of INEGI serie II is likely to be poorer than that of INEGI serie I (Mas et al., 2002b).

In the case of the National Forest Inventory (NFI) map of Mexico, a preliminary accuracy assessment was conducted immediately after map production in year 2000. A systematic sampling of the entire country was planned, but the assessment could only take place on a small portion of the planned coverage, in the Northern part of the country (Mas et al., 2002a). The assessment yielded reliability levels for a few homogeneously distributed classes, and was not designed to attend, in a cost-effective way, the high number of classes of the NFI map and their complex distribution over the entire territory.

In 2003, a research project was initiated at the Institute of Geography, UNAM, with the proposed tasks of building academic capacity for the assessment of LULC maps in Mexico and developing a framework for future accuracy assessments of the INEGI cartography. Such a framework was built in accordance with the typically available verification materials, skills and resources in Mexico. In order to implement the methodology, a pilot study was launched over a set of four distinct eco-geographical areas described in the following section. The accuracy assessment fulfilled the following desirable criteria (see section 3): 1) a probability sampling scheme (sensu Stehman & Czaplewski 1998), comprising a sampling design, a response design and the synthesis of evaluation; 2) an operational design for future INEGI map updating missions; 3) a reasonable compromise between the precision (standard error) of accuracy estimates and operational costs.

4.3 An accuracy assessment in four eco-geographical areas in Mexico

We fixed a set of eco-geographical areas (located on figure 2) that captured parts of the mega-diversity of the Mexican territory, with special focus on some of the main forest biomes (see Tables 2 and 3). They also included contrasted levels of modification of the original vegetation cover.

Two areas are located on the transversal volcanic chain and contiguous altiplano in central western Mexico. These are the closed watershed of the Cuitzeo Lake, later referred as Cuitzeo, and an area encompassing both the natural reserve of the Tancítaro peak and the Uruapan avocado production zone, later referred as Tancítaro. Both areas are included in the state of Michoacán and are covered with temperate sub-humid and tropical dry vegetation (Table 3). A third area includes the core and buffer zones of the biosphere reserve of Los Tuxtlas, in the state of Veracruz. This area is mainly characterised by tropical humid conditions although temperate humid micro-climates prevail on the relief of the two coastal volcanic chains. The fourth area corresponds to the Mexican side of the Candelaria river watershed in the state of Campeche. This area includes a portion of the Calakmul forest reserve and is mainly characterized by tropical sub-humid conditions.

The Candelaria and Tancítaro areas comprise extensive forests (of low and high levels of human management, respectively) while most of Cuitzeo and Los Tuxtlas is covered with non forested agriculture land (crop and grazing land, respectively). Apart from the informative contrast in LULC, the selection and definition of these areas were guided by the availability of reference data for independently verifying the NFI-2000 map. These reference data are detailed in table 4.

Within each eco-geographical region (stratum), the sampling design incorporated a two-stage sampling design where aerial photographic frames formed the Primary Sampling Units (PSUs), as in most regional accuracy assessments of Landsat-based maps (Wulder et al. 2006). A regular 500 m-spaced two dimensional grid (hereafter referred to as the 'second stage grid') formed the set of points, or Secondary Sampling Units (SSUs) of the second stage. Indeed, a scale criterion used during map production was to leave out polygons less than 500 meters wide.

The first stage of the sampling design consisted in the selection of two subsets of PSUs. The first subset of PSUs was obtained with a simple random selection and was used for the assessment of common classes (classes whose area fraction is above 5%, a total of 7 classes in Cuitzeo, for example). The second subset of PSUs was obtained with a proportional random selection of PSUs, and was used for the assessment of rare classes (classes whose area fraction is below 5%, a total of 14 classes in Cuitzeo, for example). In the latter selection, the probability of selection attributed to each PSU was proportional to the abundance of the rare class in that PSU, as described in Stehman et al. (2000, further discussed via personal communication); this mode of selection was retained as an appropriate way for including all scarcely distributed (or 'rare') classes (a frequent occurrence in our case), in the sample while maintaining a low complexity level of statistics (i.e. standard stratified random formulae to compute estimators of accuracy). As a compromise between the precision of the estimates and our budget for undertaking the pilot research, the number of selected PSUs approached but was maintained below one quarter of the total number of PSUs in each area. According to this scheme, the PSU selection process is made independently for each rare class and a given PSU can be potentially selected multiple times (for rare classes as well as for the common classes). This hybrid selection scheme, differentiated according to 'common' and 'rare' classes, was proposed and detailed in Couturier et al. (2007), where its potential advantages with respect to sampling designs formerly applied in the literature were evaluated.

Once the sample PSUs were selected, all points of the second stage grid included within these PSUs were assigned the attribute of their mapped land cover class. The full second stage sample consisted of the selection of 100 points (SSUs) for each class mapped in the area. For each common class, the selection was a simple random sorting of points within the second stage grid in the first subset of PSUs. For rare classes, the selection of points was obtained via proportional random sampling in the second subset of PSUs, this time with a probability inversely proportional to the abundance of the class. This mode of selection could preserve equal inclusion probabilities at the second stage within a rare class (see the option of proportional stratified random sampling advocated in Stehman et al. 2000). A sequence of ArcView and Excel-based Visual Basic simple routines, for easy and fast repeated use on vector attributes of each class, was specifically designed to perform this proportional selection at both stages.

5. Quality of detailed LULC cartography in Mexico vs. quality of international cartography

5.1 Accuracy indices of the National Forest Inventory map in Mexico

Global and per class accuracy indices are presented in table 5 for each eco-geographical area. Confusion patterns among classes were presented in error matrices by Couturier et al. (2010) and permitted a detailed study of the quality of the cartography in terms of biodiversity representation. The global accuracy indices ranged from 64 per cent (Candelaria) to 78 per cent (Los Tuxtlas). Accuracy levels were lower in forest-dominated Candelaria (64 per cent) and Tancítaro (67 per cent) areas than in nonforest-dominated Cuitzeo (75 per cent) and Los

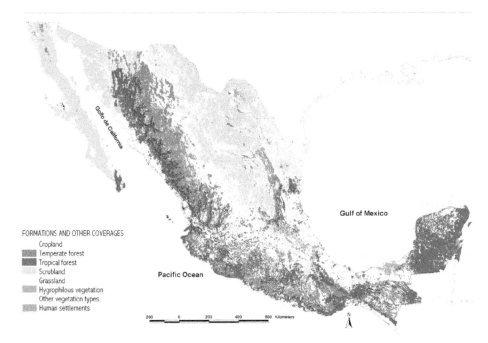

Source: Mas et al. (2002a)

Fig. 1. National Forest Inventory map of Mexico in year 2000 (NFI-2000 map)

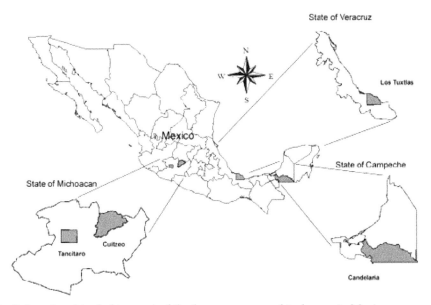

Fig. 2. Location (shaded in grey) of the four eco-geographical areas in Mexico.

Class	Name	Biome	Cuitzeo		Tancítaro		Tuxtlas		Candelaria		Total area per class (km²)
			Area frac	Area (km²)	Area frac	Area (km²)	Area frac	Area (km²)	Area frac	Area (km²)	
100	Irrigated crop	Cropland	0.1411	564.97	0.0106	13.45					578.42
110	Hygrophilous crop		0.0048	19.04							19.04
130	Cultivated grassland						0.6058	1839.75	0.1708	1908.25	3748.01
200	Perennial crop		0.0021	8.27	0.2904	367.70	0.0129	39.16			415.13
210	Annual crop		0.2356	943.14	0.0803	101.69	0.1765	535.84	0.0070	77.85	1658.51
300	Forest plantation		0.0071	28.24							28.24
410	Fir forest	Temperate forest	0.0037	14.72							14.72
420	Pine forest		0.0041	16.32	0.1658	209.99	0.0011	3.36			229.67
421	Pine forest & secondary vegetation		0.0036	14.31	0.0634	80.23	0.0011	3.37			97.90
510	Oak-pine forest		0.0958	383.34	0.1907	241.47					624.82
511	Oak-pine forest & secondary vegetation		0.0325	130.29	0.1284	162.54	0.0028	8.48			301.31
600	Oak forest		0.0232	92.88			0.0011	3.44			96.32
601	Oak forest & secondary vegetation		0.0553	221.54	0.0017	2.16	0.0041	12.49			236.20
700	Cloud forest	Tropical forest	0.0029	11.73			0.0035	10.78			22.51
800	Median/high perennial tropical forest						0.1213	368.43			368.43
801	Median/high perennial tropical forest & secondary vegetation						0.0292	88.56			88.56
820	Median/high subperennial tropical forest								0.5010	5595.31	5595.31
821	Median/high subperennial tropical forest & secondary vegetation								0.0880	982.82	982.82
830	Low subperennial tropical forest								0.1765	1971.60	1971.60
831	Low subperennial tropical forest & secondary vegetation								0.0025	27.57	27.57
920	Subtropical scrubland	Scrubland	0.0194	77.58							77.58
921	Subtropical scrubland & secondary vegetation		0.0768	307.25							307.25
1000	Mezquital		0.0004	1.51							1.51
1200	Chaparral										0.00

Code	Name	Taxonomic level									
1320	Savanna	Grassland							0.0108	120.13	120.13
1330	Induced grassland		0.1594	638.04	0.0032	4.02	0.0004	1.08	0.0039	43.65	686.80
1400	Mangrove	Hygrophilous vegetation					0.0066	20.15	0.0060	66.93	87.08
1410	Hygrophilous grassland		0.0209	83.50			0.0019	5.86	0.0225	251.24	340.60
1510	Halophilous vegetation	Other vegeta-tion types	0.0069	27.78					0.0039	43.56	71.34
1600	No apparent vegetation cover	Other cover types			0.0390	49.43	0.0007	2.11			51.54
1700	Human settlement		0.0250	100.02	0.0265	33.59	0.0065	19.82	0.0009	10.19	163.62
1800	Water		0.0796	318.75			0.0244	74.01	0.0063	70.11	462.87
	All		1.0000	**4003.23**	1.0000	**1266.28**	1.0000	**3036.69**	1.0000	**11 169.21**	**19 475.41**

Area frac: Fraction of the eco-geographical area. The 'community with alteration' taxonomic level refers to the 'sub-community' level in Palacio-Prieto et al. (2000).

Table 3. Class distribution (subcommunity and biome aggregation levels) of the NFI-2000 map in the four ecogeographical areas

Aerial photography	Data type/ interpretation	Scale/resolution	Year	Number of photographs
Cuitzeo	Prints/stereoscopic	1:37 000	1999	244
Tancítaro	Prints/stereoscopic	1:24 000	1996	152
Tuxtlas	Digital/on screen	1:75 000 / 1.5 m grain	2000 1996	12 14
Candelaria	Prints/stereoscopic	1:75 000	Jan 2000–Mar 2002	174

Table 4. Aerial photography used for the accuracy assessment of the NFI-2000 map.

Tuxtlas (78 per cent), possibly because of the higher (confusion prone) diversity of forest classes than nonforest classes in the NFI classification scheme.

For 'other cover types' ('no vegetation cover', 'water' and 'human settlement'), a high accuracy (79 per cent and above) was registered, with the only exception of 'water' in Candelaria, where water bodies are small, dispersed and often seasonal. Visually, the spectral separability of these land covers within their group and with respect to other groups is indeed among the highest on conventionally used Landsat colour composites (e.g. 342). The mangrove class also recorded high accuracy in both Candelaria and Los Tuxtlas areas where mangroves are present. Conversely, very high interconfusion within aquatic non tree vegetation covers is evident when hygrophilous grassland and halophilous vegetation are both present (Cuitzeo and Candelaria). We also found high levels of commission error in hygrophilous grassland at the expense of induced grassland in Los Tuxtlas and Candelaria. The spectral ambiguity and variability (across inundation phases) of these aquatic vegetation types is probably one of the key explanations for this observed high confusion. Former INEGI maps mostly confirm the reference data in exhibiting such errors of the NFI-2000 map. Finer trends registered for forest types and land use categories vary according to the ecogeographical area as described in Couturier et al. (2010).

By contrast with the relatively high levels of accuracy of vegetation cover with little modification (classes without 'secondary vegetation'), many errors were reported for classes

Class Code	Taxonomic name (Community with alteration)	(Biome)	Cuitzeo User's	Cuitzeo Prod-ucer's	Tancítaro User's	Tancítaro Prod-ucer's	Tuxtlas User's	Tuxtlas Prod-ucer's	Candelaria User's	Candelaria Prod-ucer's	Total area per class (km2):
100	Irrigated crop	Cropland	87	90	22	23					578.42
110	Hygrophilous crop		63	75							19.04
130	Cultivated grassland						83	90	69	78	3748.01
200	Perennial crop		99	100	86	84	57	9			415.13
210	Annual crop		71	78	87	64	52	99	75	9	1658.51
300	Forest plantation		83	33							28.24
410	Fir forest	Temperate forest	76	100							14.72
420	Pine forest		79	59	41	44	85	31			229.67
421	Pine forest & sec veg		12	5	8	44	0	-			97.90
510	Oak-Pine forest		96	92	77	67					624.82
511	Oak-Pine forest & sec veg		45	68	56	55	6	83			301.31
600	Oak forest		92	40		-	28	32			96.32
601	Oak forest & sec veg		46	95	5	100	70	82			236.20
700	Cloud forest	Tropical forest	0	-			100	100			22.51
800	Median/high perennial trop forest						92	66			368.43
801	Median/high perennial trop forest & Sec Veg						63	42			88.56
820	Median/high subperennial trop forest								70	89	5595.31
821	Median/high subperennial trop forest & Sec Veg								55	45	982.82
830	Low subperennial trop forest								52	61	1971.60
831	Low subperennial trop forest & Sec Veg								32	1	27.57
920	Sub-tropical scrubland	Scrubland	78	29							77.58
921	Sub-tropical scrubland & Sec Veg		88	63							307.25
1000	Mezquital		0	-							1.51
1200	Chaparral			-							0.00
1320	Savanna	Grassland							22	-	120.13
1330	Induced grassland		60	91	36	66	69	11	67	26	686.80
1400	Mangrove	Hygrophilous vegetation					86	99	87	96	87.08
1410	Hygrophilous grassland		47	68			53	100	70	44	340.60
1510	Halophilous vegetation	Other vegetation types	25	21					9	41	71.34
1600	No apparent vegetation	Other cover types			82	92	87	100			51.54
1700	Human settlement		100	63	97	88	92	92	80	72	163.62
1800	Water		89	92			100	98	48	96	462.87
	Total		**74.6**		**67.3**		**77.9**		**64.4**		**19475.41**

Same conventions as table 2. Trop: Tropical; Sec Veg: Secondary Vegetation; Taxonomic level *Community with alteration* refers to level *Sub-community* in Palacio-Prieto et al. (2000)

Table 5. Accuracy indices (user's and producer's) per class of the National Forest Inventory (*Community with alteration*) in the four eco-geographical areas.

of highly modified vegetation cover (classes 'with secondary vegetation'). For instance in Cuitzeo, the accuracy of sub-tropical scrubland (78%), oak-pine forest (97%), pine forest (79%) and fir forest (76%) contrast with the accuracy of highly modified oak forest (46%), highly modified pine forest (12%) and highly modified mixed forest (45%). From both the taxonomical and landscape points of view, a class of highly modified vegetation cover is close to a wide set of land use classes as well as low modification vegetation cover classes, which makes it prone to more confusions than a class of low modification vegetation cover. These low accuracy levels, however, appear as a real challenge for improving the quality of future cartography because degradation studies are an important part of forest management and biodiversity monitoring.

Region of the world	Acronym of project and year of cartography	Prevailing biotic environment	Assessment design	Number of assessed classes		Global accuracy index	Reference publication
				Forest	Total		
Mexico (4 areas)	IFN 2000	Mega-diverse	Probabilistic	19 (29)	32 (75)	64-78%	Couturier et al. (2010)
Southwest USA	GAP 2000	Temperate dry	Partial (near to roads)	18 (27)	85 (125)	61%	Lowry et al. (2007)
India	ISRO-GBP 1999	Mega-diverse	Partial (in 3 states of the country)	14	35	81%	Joshi et al. (2006)
Canada	EOSD-Forest 2000	Temperate	Probabilistic	10	18	67%	Wulder et al., (2007)
European Union	CorineLC 2000	Temperate	Probabilistic	3	22 (44)	74.8%	Buttner & Maucha (2006)
USA	NLCD 1992	Temperate	Probabilistic	3	21	46-66% (per administrative region)	Stehman et al. (2003)
Nigeria	1990	Tropical humid	Partial	3	8	74.5%	Rogers et al. (1997)
Legal Amazon, Brasil	GLC 2000	Tropical humid and dry	Partial	3	5	88%	Carreiras et al. (2006)
South Southeast Asia	TREES 2000	Mega-diverse	Probabilistic for biome level	1 (17)	4 (40)	72% (biome level)	Stibig et al. (2007)

Same conventions as table 1.

Table 6. Global accuracy indices of regional Land Use Land Cover cartography, derived from major published assessment studies in the world. The list is sorted by the number of assessed forest classes.

5.2 Comparison with other assessed international cartography

Table 6 presents the global accuracy indices found in each study listed in table 1. As a means of acknowledging the difficulty of mapping forest classes, the list in table 6 was sorted by the number of forest classes actually assessed in the study. With the exception of the GAP2000 very detailed study, the partial (non probabilistic) assessments yield higher accuracy indices (from 74.5 to 88%) than probabilistic assessments (in bold; from 46 to 74.8%). However, a partial assessment is possibly optimistically biased because it is not representative of the quality of the entire map, although it is impossible to estimate the magnitude of this bias (Stehman & Czaplewski, 1998). Among probabilistic assessments, the accuracy index in both densely forested areas (Tancítaro: 64.4% and Candelaria: 67.3%) is comparable with the results of assessments with a high amount of forest classes, en Canada (67%). Likewise, the accuracy index in areas where land use classes prevail (Cuitzeo: 74.6% and Los Tuxtlas: 77.9%) is comparable with the results of other assessment, such as the CorineLC 2000, mainly focused on land uses in Europe (74.8%)and with TREES2000 in South and Southeast Asia (72%). The accuracy indices in Cuitzeo and Los Tuxtlas, nevertheless, are higher than the range of results in other probabilistic assessments (46-66%). The NFI and the TREES 2000 cartographies have similar spatial detail (1km2 resolution) although the assessment of TREES 2000 was at biome level (only 4 assessed classes). The cartographic challenge of the NFI 2000 was greater at taxonomical detail of 'community with alteration' (32 classes).

The NFI map is also characterized by a higher taxonomic diversity than the other probabilistically assessed maps in the USA, Canada and Europe. However, the Minimum Mapping Unit (MMU) of those maps is much smaller (approximates the Landsat pixel size) than the MMU of the NFI, which in turn is a greater challenge for mapping accuracy. Considering these compensating factors (taxonomic richness but poorer spatial precision), the NFI map achieves comparable or better accuracy indices than the cited cartography, in a limited extent of the Mexican territory but in an extent that may reflect several scenarios and complexity of the national LULC.

The low accuracy registered for highly modified vegetation classes has been observed in the EOSD Canadian experience for forest covers of various density grades. Wulder et al. (2007) conclude that the highest source of errors in their map is caused by confusions among density grades. The confusion among density/ alteration classes caused by ambiguity on the Landsat imagery could be related, in the case of the NFI map, to the inclusion of the secondary vegetation in a great number of forest classes. This inclusion may be simpler and less confused in other projects such as GAP2000, TREES2000, or the NFI of year 1994 in Mexico where in spite of many forest classes, the presence of secondary vegetation is aggregated in very few classes.

A possible improvement of the detailed LULC cartography in Mexico could derive, therefore, from aggregating secondary vegetation classes into, for example, forest subtypes such as 'temperate forest with secondary vegetation' and 'tropical forest with secondary vegetation'. Such grouping could reflect a better matching of the classification system with the discrimination capacity of Landsat-like sensors in complex forest settings.

6. Conclusion

Land cover maps with detailed forest taxonomy are an essential basis for sustainable forest management at regional scale. This cartography is especially useful in highly biodiverse areas. A deforestation rate, a biodiversity conservation program or a land use change study critically depend on the quality of such cartographical datasets. Yet, for the overwhelming majority of governmental agencies in the world, the quality of the cartography is easily confounded with the spatial resolution, or temporality of the satellite imagery used in the map production process. Confusions between thematic classes on the imagery that lead to errors on the map are simply ignored, so that the derived deforestation rates, forest extent baselines, etc. are quantities without error margins and therefore these quantities lack statistical validity.

Based on a review on accuracy assessment studies in the world, this chapter first reports the occurrence of substantial errors in detailed regional land cover maps. The chapter then reports the recently developed research on the quality assessment of the LULC cartography in Mexico. A probabilistic accuracy assessment framework was developed for the first time in a mega-diverse area for taxonomically detailed maps and applied to four distinct eco-geographical areas of the Mexican NFI map of year 2000.

As a first feature of the accuracy assessment, a two-stage hybrid sampling design was applied to each of the four eco-geographical areas. Proportional stratified sampling was employed for sparsely distributed (rare) classes. This design had been fully tested and compared with existing designs in Couturier et al. (2007).

Second, with the utilization of reference maplets and GIS techniques, this research incorporated thematic and positional uncertainty as two parameters in the design, which created the possibility for a map user to evaluate the map at desired levels of positional and thematic precision. Couturier et al. (2009a) illustrated the practical usefulness of this possibility in the case of the NFI map, with landscapes composed of intricate tropical forest patches.

The accuracy of the NFI map was then compared with published error estimates of regional LULC cartographic products. We found that the quality of the NFI 2000 map (accuracy between 64% and 78%) is of international standards. This information is valuable given that the taxonomical diversity enclosed in the NFI is much higher than the currently assessed international cartography. Additionally, we found that the majority of land use classes and of low modification vegetation cover classes in the NFI are characterized by accuracy indices beyond 70%. By contrast, the NFI map registers low accuracy for highly modified vegetation cover classes. It is suggested that the quality of the cartography could be improved in the future by grouping categories containing secondary vegetation.

The assessment of the NFI 2000 cartography in four eco-geographical areas still constitutes a pilot study, confined to a limited extension, in a mega-diverse area. Since 2003, the monitoring of vegetation cover in Mexico is partly ensured using the MODIS sensor (CONAFOR, 2008), which is comparable with the SPOT-VEGETATION sensor used by Stibig et al. (2007) in Asia. We recommend the method presented here be extended to the national level for comprehensive accuracy assessment of future INEGI Serie V or vegetation cover annual maps of SEMARNAT. This method would ensure very reasonable costs and would contribute to solve the polemical discussions on the reliability of deforestation rates and land use change rates in the country.

We conclude that the work presented here sets grounds, as the first exercise of its kind, for the quantitative accuracy assessment of LULC cartography in highly bio-diverse areas. Among assets of this work is the knowledge, for the first time in a highly bio-diverse region, of the LULC quality that can be expected from the interpretation of medium resolution satellites.

7. Acknowledgments

This work was conducted under research projects 'Observatorio Territorial para la Evaluación de Amenazas y Riesgos (OTEAR)', funded by DGAPA (PAPIIT IN-307410) and

'Desarrollo de Redes para la Gestión Territorial del Corredor Biológico Mesoamericano – México', funded by CONACYT (FORDECYT 143289).

8. References

Aronoff, S. (1982). Classification Accuracy: A user approach. *Photogrammetric Engineering & Remote Sensing*, 48 (8), pp. 1299-1307.

Burrough, P.A. (1994). Accuracy and error in GIS, In: *The AGI Sourcebook for Geographic Information Systems 1995*, Green, D.R. y D Rix (Eds.), pp. 87-91, AGI, London.

Büttner, G, & G. Maucha (2006). The thematic accuracy of CORINE Land Cover 2000: Assessment using LUCAS, In: *EEA Technical Report/No7/2006*, http://reports.eea.europa.eu/ Accessed 04/2007.

Card, A. (1982). Using known map category marginal frequenties to improve estimates of thematic map accuracy. *Photogrammetric Engineering & Remote Sensing*, 48(3), pp. 431-439.

Carreiras, J., Pereira J., Campagnolo M., & Y. Shimabukuro (2006). Assessing the extent of agriculture/ pasture and secondary succession forest in the Brazilian Legal Amazon using SPOT VEGETATION data. *Remote Sensing of Environment*, 101, pp. 283-298.

Cochran, W.G. (1977). *Sampling Techniques* (3rd ed.), John Wiley and Sons, New York, 428 pp.

CONAFOR (2008). Cartografía de cobertura vegetal y usos de suelo en línea. In: *CONAFOR-SEMARNAT website*. Available from: http://www.cnf.gob.mx:81/emapas/

Congalton, R.G. (1988). Comparison of sampling scheme use in generating error matrices for assessing the accuracy of maps generated from remotely sensed data. *Photogrammetric Engineering & Remote Sensing*, 54 (5), pp. 593-600.

Congalton, R.G., & K. Green (1993). A practical look at the sources of confusion in error matrix generation. *Photogrammetric Engineering & Remote Sensing*, 59, pp. 641-644.

Couturier, S., Mas, J.-F., Vega, A., & Tapia, V. (2007). Accuracy assessment of land cover maps in sub-tropical countries: a sampling design for the Mexican National Forest Inventory map. *Online Journal of Earth Sciences*, 1(3), pp. 127-135.

Couturier, S., Vega, A., Mas, J.-F., Tapia, V., & López-Granados, E. (2008). Evaluación de confiabilidad del mapa del Inventario Forestal Nacional 2000: diseños de muestreo y caracterización difusa de paisajes. *Investigaciones Geográficas (UNAM)*, 67, pp. 20-38.

Couturier, S., Mas J.-F., Cuevas G., Benítez J., Vega-Guzmán A., & Coria-Tapia V. (2009a). An accuracy index with positional and thematic fuzzy bounds for land-use/ land-cover maps. *Photogrammetric Engineering & Remote Sensing*, 75 (7), pp. 789-805.

Couturier, S., Gastellu-Etchegorry, J.-P., Patiño, P., & Martin, E. (2009b). A model-based performance test for forest classifiers on remote sensing imagery. *Forest Ecology and Management*, 257, pp. 23-37.

Couturier, S., Mas J.F., López E., Benítez J., Coria-Tapia V., & A. Vega-Guzmán (2010). Accuracy Assessment of the Mexican National Forest Inventory map: a study in four eco-geographical areas. *Singapore Journal of Tropical Geography*, 31 (2), pp. 163-179.

Equihua, M. (1990). Fuzzy clustering of ecological data. *Journal of Ecology*, 78, pp. 519-534.

Equihua, M. (1991). Análisis de la vegetación empleando la teoría de conjuntos difusos como base conceptual. *Acta Botánica Mexicana*, 15, pp. 1-16.

Fisher, P. & S. Pathirana (1990). The evaluation of fuzzy membership of land cover classes in the suburban zone. *Remote Sensing of Environment*, 34, pp. 121-132.

Fitzpatrick-Lins, K. (1981). Comparison of sampling procedures and data analysis for a land-use and land-cover map. *Photogrammetric Engineering & Remote Sensing*, 47, pp. 343-351.

Foody, G.M. (1992). A fuzzy sets approach to the representation of vegetation continuum from remotely sensed imagery: an example from lowland heath. *Photogrammetric Engineering & Remote Sensing*, 55, pp. 221-225.

Foody, G.M. (2002). Status of land cover classification accuracy assessment. *Remote Sensing of Environment*, 80, pp. 185-201.

Gopal S., & C. E. Woodcock (1994). Accuracy of Thematic Maps using fuzzy sets I: Theory and methods. *Photogrammetric Engineering & Remote Sensing*, 58, pp. 35-46.

Green, D.R., & W. Hartley (2000). Integrating photo-interpretation and GIS for vegetation mapping: some issues of error, In:*Vegetation Mapping from Patch to Planet* (Alexander, R. and A.C. Millington, editors), pp. 103-134, John Wiley & Sons Ltd.

Haack, S. (1996). *Deviant logic, fuzzy logic*. University of Chicago Press, Chicago, 146pp.

Hagen, A. (2003). Fuzzy set approach to assessing similarity of categorical maps. *International Journal of Geographical Information Science*, 17 (3), pp. 235-249.

INEGI (1980). *Sistema de Clasificación de Tipos de Agricultura y Tipos de Vegetación de México para la Carta de Uso del Suelo y Vegetación del INEGI, escala 1:125 000*. Instituto Nacional de Estadística, Geografía e Informática, Aguascalientes, Ags, México.

Joshi, P. K., P. S. Roy, S. Singh, S. Agrawal & D. Yadav (2006). Vegetation cover mapping in India using multi-temporal IRS Wide Field Sensor (WiFS) data. *Remote Sensing of Environment*, 103, pp. 190-202.

Laba, M., Gregory SK, Braden J et al. (2002). Conventional and fuzzy accuracy assessment of the New York Gap Analysis Project land cover map. *Remote Sensing of Environment*, 81, pp. 443-455.

Lowry, J., R. D. Ramsey, K. Thomas et al. (2007). Mapping moderate-scale land-cover over very large geographic areas within a collaborative framework: A case study of the Southwest Regional Gap Analysis Project (SWReGAP). *Remote Sensing of Environment*, 108 pp. 59-73.

Mas J.-F. & I. Ramírez (1996). Comparison of land use classifications obtained by visual interpretation and digital processing. *ITC Journal*, 3(4), pp. 278-283.

Mas J.-F, Velázquez A, Palacio-Prieto JL, Bocco G, Peralta A, & Prado J. (2002a). Assessing forest resources in Mexico: Wall-to-wall land use/ cover mapping. *Photogrammetric Engineering & Remote Sensing*, 68 (10), pp. 966-969.

Mas J.-F., A. Velázquez, J.R. Díaz, R. Mayorga, C. Alcántara, R. Castro, & T. Fernández (2002b). Assessing land use/cover change in Mexico, *Proceedings of the 29th International Symposium on Remote Sensing of Environment* (CD), Buenos Aires, Argentina, 8-12/04/2002.

Palacio-Prieto JL, Bocco G, Velázquez A, Mas JF, Takaki-Takaki F, Victoria A, Luna-González L et al. (2000). La condición actual de los recursos forestales en México: resultados del Inventario Forestal Nacional 2000, *Investigaciones Geográficas (UNAM)*, 43, pp. 183-202, ISSN: 0188-4611

Pontius R.G., & M.L. Cheuk (2006). A generalized cross-tabulation matrix to compare soft-classified maps at multiple resolutions. *International Journal of Geographical Information Science*, 20 (1), pp. 1-30.

Remmel T.K., & F. Csillag (2006). Mutual information spectra for comparing categorical maps. *International Journal of Remote Sensing*, 27 (7), pp. 1425-1452.

Rogers, D. J., S. I. Hay, M. J. Packer & GR. Wint (1997). Mapping land-cover over large areas using multispectral data derived from the NOAA-AVHRR: a case study of Nigeria. *International Journal of Remote Sensing*, 18 (15), pp. 3297-3303.

Rzedowski J. (1978). *Vegetación de México*, Limusa, México City

Särndal, C.E., Swensson V., & J. Wretman (1992). *Model-assisted survey sampling*, Springer-Verlag, New-York

Snedecor, G.W., & W.F. Cochran (1967). *Statistical methods*, State University Press, Ames, Iowa, 728 pp.

Stehman S.V., & R.L. Czaplewski (1998). Design and analysis for thematic map accuracy assessment: fundamental principles. *Remote Sensing of Environment*, 64, pp. 331-344.

Stehman, S.V., Wickham J.D., Yang L., & J.H. Smith (2000). Assessing the Accuracy of Large-Area Land Cover Maps: Experiences from the Multi-Resolution Land-Cover Characteristics (MRLC) Project, *4th International Symposium on Spatial Accuracy Assessment in Natural Resources and Environmental Sciences (Accuracy 2000)*, Amsterdam, pp. 601-608.

Stehman, S.V. (2001). Statistical rigor and practical utility in thematic map accuracy assessment. *Photogrammetric Engineering & Remote Sensing*, 67, pp. 727-734.

Stehman, S.V., Wickham JD, Smith JH, & Yang L (2003). Thematic accuracy of the 1992 National Land-Cover Data for the eastern United-States: Statistical methodology and regional results. *Remote Sensing of Environment*, 86, pp. 500-516.

Stibig, H. J., A. S. Belward, P. S. Roy, U. Rosalina-Wasrin et al. (2007). A land-cover map for South and Southeast Asia derived from SPOT- VEGETATION data. *Journal of Biogeography*, 34, pp. 625-637

Velázquez, A., Mas J.-F., Díaz J.R., Mayorga-Saucedo R., Palacio-Prieto J.L., Bocco G., Gómez-Rodríguez G. et al. (2001). El Inventario Forestal Nacional 2000: Potencial de uso y alcances. *Ciencias*, 64, pp. 13-19.

Velázquez, A., Mas J.-F., Díaz J.R., Mayorga-Saucedo R., Alcántara P.C., Castro R., Fernández T., Bocco G., Escurra E., & J.L. Palacio-Prieto (2002). Patrones y tasas de cambio de uso del suelo en México. *Gaceta Ecológica*, INE-SEMARNAT, 62, pp. 21-37.

Visser, H., & T. de Nijs (2006). The map comparison kit. *Environmental Modelling & Software*, 21, pp. 346-358.

Wang, F., (1990). Improving remote sensing image analysis through fuzzy information representation. *Photogrammetric Engineering and Remote Sensing*, 56 (9), pp. 1163-1168.

Wickham, J.D., Stehman S.V., Smith J.H., & L. Yang (2004). Thematic accuracy of the 1992 National Land-Cover Data for the western United-States. *Remote Sensing of Environment*, 91, pp. 452-468.

Wulder, M.A., Franklin S.F., White J.C., Linke J., & S. Magnussen (2006), An accuracy assessment framework for large-area land cover classification products derived from medium-resolution satellite data. *International Journal of Remote Sensing*, 27(4), pp. 663-68.

Wulder, M. A., J. C. White, S. Magnussen, & S. McDonald (2007). Validation of a large area land cover product using purpose-acquired airborne video. *Remote Sensing of Environment* 106 pp. 480-491.

Zadeh, L. (1965), Fuzzy sets. *Information and control*, 8, pp. 338-353.

Zhu Z., Yang L., Stehman S.V., & R.L. Czaplewski (2000), Accuracy Assessment for the U.S. Geological Survey Regional Land-Cover Mapping Program: New York and New Jersey Region. *Photogrammetric Engineering & Remote Sensing*, 66, pp. 1425-1435.

3

Remote Monitoring for Forest Management in the Brazilian Amazon

André Monteiro and Carlos Souza Jr.
Amazon Institute of People and The Environment-Imazon
Brazil

1. Introduction

Timber harvesting is an important economic activity in the Brazilian Amazon. In 2009, the timber industry produced 5.8 million cubic meters of logwood and generated US$ 2.5 billion in gross income along with 203,705 direct and indirect jobs (Pereira et al., 2010). Logging in the region is predominantly predatory, and is commonly known as Conventional. Only a small proportion occurs in a managed fashion (planned), known as Reduced Impact Logging (RIL) (Asner et al., 2002; Gerwing, 2002; Pereira Jr. et al., 2002; Veríssimo et al., 1992). In the conventional method activities are not planned (opening of roads and log decks[1], tree felling and log skidding), while with RIL planned management techniques are applied at all stages of harvesting (Amaral et al., 1998).

The two methods cause impacts ranging from low to severe on the structure and composition of the remaining forest (Gerwing, 2002; John et al., 1996; Pereira Jr. et al., 2002). However, the impacts of predatory logging are two times greater than those of managed logging (John et al., 1996). Among the main impacts are: greater reduction in living aboveground biomass (Gerwing, 2002; Monteiro et al., 2004), risk of extinction for high-value timber species (Martini et al., 1994), greater susceptibility to forest fires (Holdsworth & Uhl, 1997), increase of vines and pioneer vegetation (Gerwing, 2002; Monteiro et al., 2004) and substantial reduction in carbon stocks (Gerwing, 2002; Putz et al., 2008).

The impact of timber harvesting can be described by means of forest inventories carried out in the field, with which it is possible to evaluate the structure and composition of the remaining forest (Gerwing, 2002; John et al., 1996; Monteiro et al., 2004). Another method employed is remote sensing, which has advanced over the last decade. In the Amazon, there have been successful tests with satellite images to detect and quantify forest degradation brought about by logging activities in the region (Asner et al., 2005; Matricardi et al., 2007; Souza Jr. et al., 2005). Images with moderate spatial resolution, such as Landsat (30 m) and Spot (20 m), have been used to detect types of logging, damages to the canopy and roads and log decks for harvesting (Asner et al., 2002; Matricardi et al., 2007; Souza Jr. & Roberts, 2005). As for images with high spatial resolution, such as Ikonos (1 to 4 m), they are capable of detecting smaller features of logging, such as small clearings (Read et al., 2003), as well as making it possible to determine the size of log decks and width of roads (Monteiro et al.,

---

[1] Clearings (500 m²) opened in the forest for storing timber.

2007). The use of remote sensing for monitoring forest management plans is of great importance for the Brazilian Amazon, given that logging activities are predominantly predatory and occur in extensive areas that are difficult to access.

Recent studies have shown how to integrate data extracted from satellite images with biomass data collected in the field, which makes it possible to estimate the loss of biomass in the forest submitted to different levels of forest degradation (Asner et al., 2002; Pereira Jr. et al., 2002; Souza Jr. et al., 2009). Our research has made advances in applying those techniques to assess the intensity and quality of logging (Monteiro et al., 2009).

In this chapter, we demonstrate how the impacts of timber harvesting can be characterized by means of forest inventories and combined with satellite images to monitor extensive areas. We also present the remote sensing techniques utilized for detecting, mapping and monitoring logging activities. Finally, we present the results of our system for monitoring forest management plans, applied in Pará and Mato Grosso, the two largest timber-producing States in the Amazon, which respectively account for 44% and 34% of the total produced in 2009 (Pereira et al., 2010).

2. Logging impact characterization based on field surveys

2.1 Change in structure and composition as a result of forest degradation

Characterization of the impacts of logging in the field is done by means of forest inventories. To do this, transects or plots are established in the forest to quantify the damages to its structure and composition in terms of soil cover, canopy cover and aboveground live biomass (Gerwing, 2002; John et al., 1996; Monteiro et al., 2004).

In the method developed by Gerwing (2002) 10 m x 500 m transects are opened, in which all individual trees with DBH (Diameter at Breast Height) ≥ 10 cm are sampled. In 10 m x 10 m sub-parcels, located at 50 m intervals along the central line of the transect, all individuals with DBH ≤ 10 cm are sampled. In those sub-parcels the soil cover is assessed, with the percentages of intact soil, soil with residues and disturbed soils being recorded; as well as the canopy cover, with four readings in a spherical densiometer, at 90° intervals, every 50 m along the central axis of the transect (Figure 1).

Additionally, the live biomass above the ground in each transect is estimated, adding together the weight of dry matter from different forest components using alometric equations available in the literature (Table 1). The estimate of biomass for trees < 10 cm is done by multiplying the number of stems in each diameter class by the biomass corresponding to the arithmetic average of the diameter for each class.

The forest inventory was carried out in 55 transects, including 11 in intact forest (reference) and 44 in forests in different classes of degradation due to different log harvesting methods. It was done in the Paragominas and Santarém regions, in the State of Pará, in Sinop, in Mato Grosso, and in Itacoatiara, in Amazonas (Figure 2). Below is a description of intact forest and forests in different classes of degradation according to Gerwing (2002):

i.   Intact forest: mature forest (> 40 years) without disturbance, dominated by shade-tolerant species.

ii.   Logged without mechanization (Traditional logging): forest logged without the use of skidder tractors, that is, without impact from construction of logging infrastructure: log decks, roads and skidder trails.

iii. Managed logging (Reduced Impact Logging-RIL): forest logged selectively following planning of harvesting activities: forest inventory, opening of decks and roads, felling and skidding of trees and transport of logs.

iv. Conventional logging: forest logged selectively and not following planning of the activities mentioned above. Log decks, roads and skidder trails are opened causing severe damage to the forest.

v. Logged and burned: forest logged selectively, without planning, followed by burning.

vi. Logged intensely and burned: forest logged selectively in a conventional manner more than once, and later burned.

vii. Burned: forest burned without having been logged.

To evaluate differences between the variables in the degradation classes we employed the analysis of variance (ANOVA with Type III Sum of Squares) followed by the Tukey's HSD post-hoc test with an individual error rate of 0.05% and with an overall significance of 0.08% using the R program (R *Development Core Team*, 2010).

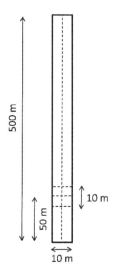

Fig. 1. Transect layout of the forest inventories.

Species group	Regression equation	Source
Forest tree species ≥ 10 cm DBH	$DW= 0.465(DBH)^{2.202}$ $DW= 0.6*4.06(DBH)^{1.76}$	Overman et al. (1994) Higuchi & Carvalho (1994)
< 10 cm DBH	$\log(DW)= 0.85+2.57 \log(DBH)$	J. Gerwing (data not published)
Pioneer tree species Cecropia sp. Other sp.	$\ln(DW)= -2.512+2.426 \ln(DBH)$ $\ln(DW)= -1.997+2.413 \ln(DBH)$	Nelson et al. (1999) Nelson et al. (1999)

Table 1. Regression equations used to determine the dry weights (kg) of various forest components based on their diameters (cm).

In the subsections below, we present the results of characterizing forest degradation for the classes described above. That information is later combined with remote sensing data to evaluate the intensity and quality of logging. In the field we quantified forest degradation related to soil disturbance (intact vegetation, residues and disturbed soil), canopy cover and aboveground biomass because those indicators present a direct relation with remote sensing data (Souza Jr. et al., 2009).

Fig. 2. Location of the forest transects sites.

2.1.1 Soil disturbance and canopy cover

The evaluation of soil disturbance (intact vegetation, residues and disturbed soil) and canopy cover in the field is crucially important, since those results directly influence the results of the satellite images. The greater the soil disturbance and the smaller the canopy cover of the degraded forest, the greater will be the signal for this damage in the image. Our results show that the area of intact vegetation was smaller in the classes with greater degradation. The smallest percentage of intact vegetation was observed in the intensely logged and burned class (4%), followed by burned forest (22%), with these presenting a significant difference in relation the intact forest and to the classes with less degradation. The quantity of residues in the soil was greater in the logged and burned forest (26%), followed by the burned forest (25%), however no significant differences were found between these classes and intact forest. The area of disturbed forest was greater in the intensely logged and burned forest (96%) and the burned forest (53%), presenting a significant difference in relation to the intact forest and the other degradation classes. The logged and burned class presented the lowest canopy cover (75%), with a significant difference in relation to the intact forest and to the classes with less degradation (Table 2).

2.1.2 Change in the live aboveground biomass

The live biomass aboveground was less in the forest degradation classes compared to the biomass in intact forest; however, no significant differences were found between them. Among individuals with DBH ≥ 10 cm, the logged and burned and intensely logged classes presented 36% lower biomass than the intact forest, followed by the burned class (18%) (Table 2). The lowest biomass for individuals with DBH < 10 cm was also observed in the logged and burned (44%) and intensely logged and burned (11%) classes (Table 2). The biomass in individuals with DBH ≥ 10 cm decreased with increasing degradation. The variation in biomass for individuals with DBH < 10 cm seems to be related to the incidence of pioneer species that tolerate moderate levels of degradation (Gerwing, 2002; Monteiro et al., 2004). The greatest loss of biomass is not related only to the greatest forest degradation. The distance from the first degradation may also influence a greater reduction in biomass (Gerwing, 2002; Monteiro et al., 2004). For example, data collection in the logged and burned forest (biomass for individuals ≥ 10 cm and < 10 cm = 232 t ha⁻¹ and 5 t ha⁻¹, respectively) occurred approximately 2.5 years after the first degradation event, while in the intensely logged and burned forest (biomass for individuals ≥ 10 cm and < 10 cm = 234 t ha⁻¹ and 8 t ha⁻¹, respectively), it occurred 16 years after the first degradation event.

	(a) Intact (n=11)	(b) Non-mechanized logging (n=9)	(c) Managed logging (n=14)	(d) Conventional logging (n=8)	(e) Logged and burned (n=4)	(f) Heavily logged e burned (n=5)	(g) Burned (n=4)
Ground cover (total area (%))							
Intact vegetation	95 (6) a	87 (22) b	96 (20) c	92 (15) ac	75 (7) d	4 (5) ef	22 (19) f
Woody debris	5 (6) a	23 (23) a	23 (12) a	13 (11) a	26 (10) a	0 a	25 (44) a
Disturbed soil	0 a	3 (4) a	10 (9) a	13 (11) a	7 (6) a	96 (5) b	53 (49) c
Canopy cover (%)	95 (3) a	87 (9) b	96 (2) a	92 (6) a	75 (8) c	86 (2) a	88 (1) d
Aboveground live biomass (t ha⁻¹)							
Live trees ≥ 10 cm DBH	365 (50) a	347 (33) a	342 (52) a	321 (12) a	232 (11) a	234 (23) a	299 (14) a
Live trees < 10 cm DBH	9 (2) a	10 (2) a	9 (1) a	11 (1) a	5 a	8 a	9 a

[*] Means presented with standard deviation noted parenthetically. In the ground cover, canopy cover and biomass values, different forest class letters denote significant differences among stand classes at $P<0.05$ utilizing Tukey's HSD post-hoc test, with a global significance level of 0.8

Table 2. Comparison of ground cover, canopy cover and biomass among intact forest and degraded forest in the States of Pará, Mato Grosso and Amazonas in Brazil[*]

3. Remote sensing techniques to enhance and detect timber harvesting

Moderate satellite imagery such as Landsat Thematic Mapper (30-meters pixel size) and Spot Multispectral (20 meters) has been used to detect and map the impacts of selective logging. However, the complex mixture of dead and live vegetation, shadowing and soils found throughout forest environments impose challenges to revealing these impacts, requiring advanced remote sensing techniques (Asner et al., 2005; Souza Jr. et al., 2005).

From the satellite vantage point, forest damage caused by logging seems to disappear within three years or less, making detection of previously logged forest (> 1 year) very challenging (Souza Jr. et al. 2009; Stone & Lefebvre, 1998). Remote sensing studies on logging in the Brazilian Amazon found that Landsat reflectance data have high spectral ambiguity for distinguishing logged forest from intact forest (Asner et al., 2002, Souza Jr. et al., 2005). Vegetation indices (Souza et al. 2005a; Stone & Lefebvre, 1998) and texture filters (Asner et al., 2002) also showed a limited capability for detecting logging. Improving the spatial resolution of reflectance data can help; 1-4 m resolution Ikonos satellite data can readily detect forest canopy structure and canopy damage caused by selective logging (Asner et al. 2002; Read et al., 2003; Souza Jr. & Roberts, 2005). However, the high cost of these images, and additional computational challenges in extracting information, requiring a combination of object-oriented classification with spectral information, severely limit the operational use of Ikonos and similar imagery.

Over the last two decades, the Brazilian Amazon has been a great laboratory for testing remote sensing techniques to detect and map forest impacts of selective logging (Asner et al., 2005; Matricardi et al., 2001; Read et al., 2003; Souza Jr. & Barreto, 2000; Souza Jr. et al., 2005; Stone & Lefebvre, 1998;). These techniques differ in terms of mapping objectives, image processing techniques, geographic extent, and overall accuracy. In terms of mapping objective, some image processing algorithms were proposed for the total logged area, including roads, log landings, forest canopy damaged and undisturbed forest islands, while others were focused only on the mapping of forest canopy damage. Techniques to map total logged area were based on visual interpretation (e.g., Matricardi et al., 2001; Stone & Lefebvre, 1998), combination of automated detections of log landings with buffer applications defined by logging extraction reach (Monteiro et al., 2003; Souza Jr. & Barreto, 2000), and textural filtering (Matricardi et al, 2007). More automated techniques are mostly based on SMA (spectral mixture analysis) approaches combined with spatial pattern recognition algorithms (Asner et al., 2005; Souza Jr. et al., 2005). Finally, image segmentation has been applied to very high spatial resolution imagery (Hurtt et al., 2003). Landsat images are the ones most used in the studies and in operational systems in the Brazilian Amazon.

Some research has shown that the detection of logging at moderate spatial resolution is best accomplished at the sub-pixel scale using SMA (Box 1). Images obtained with SMA that show detailed fractional cover of soils, non-photosynthetic vegetation (NPV) and green vegetation (GV) enhance our ability to detect logging infrastructure and canopy damage. For example, log landings and logging roads have higher levels of exposed bare soil with detection facilitated by Soil Fraction (Souza Jr. & Barreto, 2000). The brown vegetation component, including trunks and tree branches, increases with canopy damage, making NPV fraction useful for detecting this type of area (Souza Jr. et al., 2003; Cochrane & Souza Jr., 1998) and the green vegetation (GV) fraction is sensitive to canopy gaps (Asner et al., 2004).

A novel spectral index combining the information from these fractions, the Normalized Difference Fraction Index (NDFI) (Souza, Jr. et al., 2005), was developed to augment the detection of logging impacts. NDFI is computed as:

$$NDFI = \frac{GV_{Shade} - (NPV + Soil)}{GV_{Shade} + NPV + Soil} \qquad (1)$$

where GV_{shade} is the shade-normalized GV fraction given by,

$$GV_{Shade} = \frac{GV}{100 - Shade} \tag{2}$$

NDFI values range from -1 to 1. For intact forests, NDFI values are expected to be high (i.e., about 1) due to the combination of high GV shade (i.e., high GV and canopy Shade) and low NPV and Soil values. As forest becomes degraded, the NPV and Soil fractions are expected to increase, lowering NDFI values relative to intact forest (Souza Jr. et al., 2005). Canopy damage detection caused by forest degradation induced by factors such as logging and forest fires can be detected with Landsat images within a year of the degradation event with 90.4% overall accuracy (i.e., for three land cover classes, Non-Forest, Forest and Canopy Damage) (Souza Jr. et al., 2005).

The reflectance data obtained from Landsat data of each pixel can be decomposed into endmember fractions, which are purest component materials that are expected to be found within the image pixels. For the purpose of detecting forest degradation, we modeled the reflectance pixel in terms of GV (green vegetation), NPV (non-photosynthetic vegetation), Soil and Shade through Spectral Mixture Analysis – SMA (Adams et al., 1993). The SMA model assumes that the image spectra are formed by a linear combination of n pure spectra, such that:

$$R_b = \sum_{i=1}^{n} Fi\, R_{i,b} + \varepsilon_b \tag{1}$$

for

$$\sum_{i=1}^{n} F_i = 1 \tag{2}$$

where R_b is the reflectance in band b, $R_{i,b}$ is the reflectance for endmember i, in band b, F_i the fraction of endmember i, and ε_b is the residual error for each band. The SMA model error is estimated for each image pixel by computing the **RMS** error, given by:

$$RMS = [n^{-1} \sum_{b=1}^{n} \varepsilon_b]^{1/2} \tag{3}$$

Identifying the correct endmembers is a crucial step in SMA model. To avoid subjectiveness in this process, we have built a generic endmember spectral library (Figure 3) as described in Souza Jr. et al. (2005).

The following steps are used to evaluate SMA results :
1. Fraction images are evaluated and interpreted in terms of field context and spatial distribution. For example, high Soil fraction values are expected in roads and log landings and high NPV in forest areas with canopy damage;
2. Fraction values should have physically meaningful results (i.e., fractions ranging from zero to 100%). Histogram analysis of fraction values can be performed to evaluate this requirement.

3. Fraction values must be consistent over time for invariant targets, i.e., that intact forest not subject to phenological changes must have similar values over time.

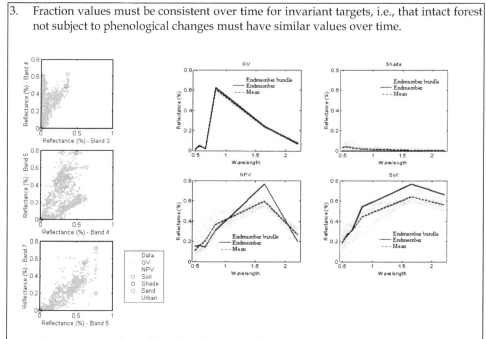

Fig. 3. Image scatter-plots of Landsat bands in reflectance space and the spectral curves of GV, Shade, NPV and Soil (source: Souza Jr. et al., 2005).

Box 1. Spectral Mixture Analysis (SMA)

4. Integrating field and remote sensing data

Assessment of the quality of timber harvesting has traditionally been done through measuring damages to the forest, e.g. quantification of the opening of log decks, logging roads and openings resulting from felling trees; and the density and biomass for remaining individuals (Gerwing, 2002; Pcreira Jr. et al., 2002; Veríssimo et al., 1992). However, field surveys are expensive and lengthy, especially for extensive areas such as the Amazon. Recent studies have shown that it is possible to infer the quality of timber harvesting through satellite images that are calibrated with indicators of damages measured in the field, allowing greater speed, reduction of costs and monitoring of extensive areas. Using satellite images such as Landsat and Spot it is possible to evaluate the quality of logging activities based on mapping of roads, log decks and damages to the forest canopy (Monteiro & Souza Jr., 2006; Monteiro et al., 2009). We present below the items evaluated and the respective indicators for monitoring timber harvesting and the results of its application in order to qualify its impacts.

4.1 Roads and log decks
For the roads and log decks we evaluated the following indicators: the density of log decks and roads; the distance between secondary roads and between log decks; and spatial

distribution of log decks and roads. Those indicators were tested in 43 logging areas located in regions of Pará, Mato Grosso and Amazonas. The results were validated with measurements of the same indicators in the field, in areas of conventional (predatory) logging and managed logging in the Paragominas (PA) and Sinop (MT) regions (Monteiro & Souza Jr., 2006).

To do this we used Landsat 5 TM satellite images with 30 meters of spatial resolution. We first applied geometric and atmospheric correction to those images. Next, we obtained fraction images of vegetation, soils and NPV (non-photosynthetically active vegetation), based on the spectral mixture model followed by NDFI (Normalized Difference Fraction Image) to highlight the scars caused by logging (Souza Jr. et al., 2005). Finally, we digitalized the log decks and roads in the NDFI image and inferred the density of log decks and roads and the distances between log decks and between roads. Additionally, we classified the spatial distribution of log decks and roads as systematic and non-systematic. Systematic distribution is characterized by rectilinear and parallel roads and log decks regularly distributed along the roads, while non-systematic distribution is defined by sinuous roads with log decks interlinked by their segments.

The results of evaluating the indicators presented an average density of 16 meters/hectare for roads and 3/100 hectares for log decks. The average distance between roads was 623 meters, and between log decks it was 484 meters (Table 3). Conventional logging presented a higher density of log decks and roads compared to logging with forest management (Johns et al., 1996). As for the distance between secondary roads and between log decks, they are smaller in logging with forest management compared to the distances between secondary roads in conventional logging (Monteiro, 2005).

As for the spatial distribution of log decks and roads, the majority of areas evaluated that were logged using forest management presented a non-systematic distribution of log decks (60%) and roads (58%), which indicates low quality in planning that infrastructure (Monteiro & Souza Jr., 2006).

	Region	Logging type	Density*		Distance*	
			Road (m/ha)	Log deck (n/100 ha)	Secondary roads (m)	Log landing (m)
IMAGE	Pará, Mato Grosso and Amazonas	Managed	16 (5)	3 (2)	623 (232)	484 (148)
FIELD	Paragominas (PA)	Managed	23 (4)	8 (1)	469 (30)	260 (74)
		Conventional	36 (6)	15 (2)	513 (38)	301 (263)
	Sinop (MT)	Managed	32 (11)	7 (4)	455 (24)	347 (126)
		Conventional	19 (3)	1 (2)	508 (43)	512 (44)

*Mean density and distance with standard deviation within brackets.

Table 3. Comparison between the indicators (mean) measured in the images from Pará, Mato Grosso and Amazonas regions and those measures in the field in Paragominas and Sinop (Monteiro & Souza Jr., 2006).

4.2 Forest canopy

We evaluated the indicator of damages to the canopy caused by logging operations. To do that we utilized NDFI images to evaluate the area of forest affected. First, we delimited the logging area visible in the NDFI image by means of visual interpretation. Next, we selected around five samples from 100 in the NDFI image to represent logging and extracted the average values of those samples. The samples were composed of a mosaic of environments (forest, log decks, roads, skidder trails and clearings caused by felled trees).

The quality of logging is determined using thresholds obtained in the NDFI image and calibrated using field data (Monteiro et al., 2008), so that: NDFI≤ 0.84 represents low quality timber harvesting (predatory logging); NDFI= 0.85-0.89, intermediate quality harvesting (there was an attempt at adopting management, but the configuration of roads, log decks and clearings reveals serious problems with execution); and NDFI≥ 0,90, good quality harvesting (the configuration of roads, log decks and clearings is in conformity with the techniques recommended by forest management (Figure 4).

This method was tested in the States of Pará and Mato Grosso, the main timber producers, responsible respectively for 44% and 34% of the total produced in 2009 in the Brazilian Amazon (Pereira et al., 2010). We evaluated 156,731 and 177,625 hectares respectively of areas undergoing timber harvesting in the two States. In Pará, 21% of that total presented logging of good quality, 54% showed intermediate quality and 25% showed low quality (Figure 5). In Mato Grosso, only 9% of logging was of good quality, 55% showed intermediate quality and 36% showed low quality (Figure 5). In the images, the log decks appear as yellow points; and the roads as light green lines. In the areas with logging of good quality, we observed the low impact on the canopy as light green patches in the images. The medium impact on the canopy observed in areas with intermediate quality appears as intense light green patches. In low quality harvesting, the log decks and roads are mixed, with the high impact on the canopy and appearing as more intense patches, varying from light green to yellow) (Figure 4). The high percentage of areas harvested in Pará and Mato Grosso with intermediate and low quality indicates a low level of adoption of forest management. This may also point to technical deficiency among company forest management technicians.

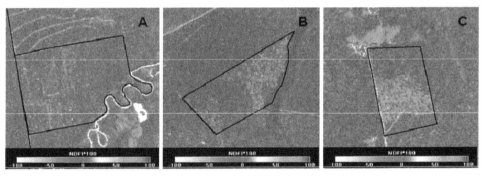

Fig. 4. Forest management of good (A), intermediate (B) and low (C) quality according to NDFI images.

To validate results of our assessment of the quality of timber harvesting as seen in satellite images, we went to the field to quantify it. To do this, we evaluated and scored impacts of logging resulting from opening of log decks and roads, felling trees and damages to remaining trees. We verified that the greater the impact, the lower the quality of harvesting and vice-versa. We thus attributed a score (from 0 to 4) and a corresponding classification (low, intermediate and good), in which: score <2 = low quality; score 2-<3 = intermediate quality; and score 3-4 = good quality. In table 4 we present the results of that validation. However, in the samples validated in the field we did not have cases of low quality management, despite having detected this standard of quality in the images. The quality standards were correctly classified in 86% and 58% of cases, as intermediate and good quality respectively (Table 4). The cases in which results in the images were different from those in the field may be related to the fact that the area evaluated in the field was geographically not the same area evaluated in the image. In the image we sampled the forest management area that visually was the most disturbed; however, because of the difficulty in accessing that area in the field, we had to evaluate another area geographically closed to the real area. With this, we verified that the quality of forest management can vary within the same licensed area, confirming the importance of monitoring forest management by satellite as a planning tool in enforcement campaigns by environmental agencies.

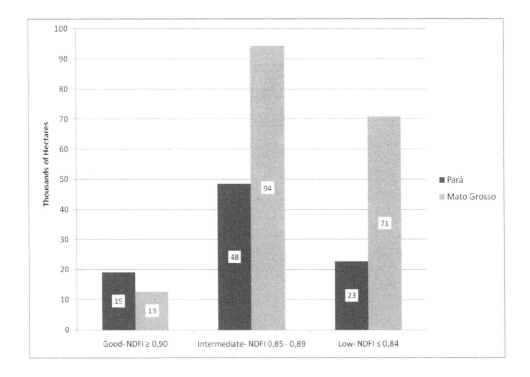

Fig. 5. Quality (in hectares) of timber harvesting in management plans in the State of Pará and Mato Grosso.

Forest Management Sample	Quality in the image		Quality in the field	
	Classification	NDFI	Classification	Scoring
1	intermediate	0.86	intermediate	2.77
2	intermediate	0.89	intermediate	2.66
3	intermediate	0.86	good	3.43
4	good	0.90	good	3.18
5	intermediate	0.85	intermediate	2.42
6	intermediate	0.88	intermediate	2.26
7	good	0.90	good	3.22
8	good	0.90	good	3.00
9	good	0.90	intermediate	2.54
10	intermediate	0.88	intermediate	2.72
11	good	0.91	intermediate	2.81
12	good	0.90	good	3.09
13	good	0.91	good	3.09
14	good	0.90	good	3.36
15	good	0.90	intermediate	2.81
16	good	0.91	good	3.45
17	good	0.90	intermediate	2.45
18	good	0.90	intermediate	2.45
19	intermediate	0.89	intermediate	2.45
20	good	0.91	intermediate	2.90
21	good	0.91	good	3.27
22	good	0.91	good	3.18
23	good	0.91	intermediate	2.45
24	good	0.90	intermediate	2.54
25	good	0.91	good	3.09
26	good	0.90	good	3.27

Table 4. Comparison of forest management quality obtained in the image and obtained in the field (Monteiro et al., 2011).

4.3 Forest biomass

We evaluated the loss of forest biomass indicator in the areas submitted to forest degradation. To do this, we first obtained an NDFI image to quantify forest degradation (Souza Jr. et al., 2005). Next, we integrated that information with the forest biomass data collected in the field (See section 2.1.2).

We observed that the NDFI value in the image diminishes with the increase in forest degradation. This means that the lower the biomass, the more degraded the forest; in other words, there is a high negative correlation between the biomass values quantified in the field with the NDFI values of the forest degradation classes (Figure 6) (Souza Jr. et al., 2009). However, that negative correlation is only observed when the NDFI image is from the same year as the occurrence of the degradation event, since beginning in the following year, the degradation signal diminishes (Souza Jr. et al., 2009).

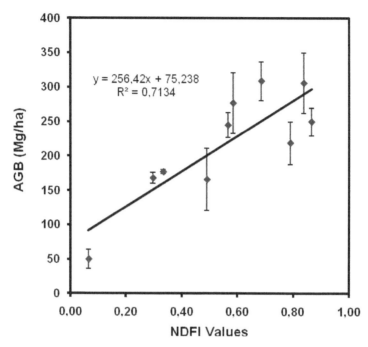

Fig. 6. Relationship between Aboveground Biomass- AGB and NDFI values for degraded forest of Paragominas and Sinop (Souza Jr. et al., 2009).

5. Applying remote sensing to monitor forest management in the Amazon

In the subsections below we present the results of remote monitoring in the timber harvesting areas of the States of Pará and Mato Grosso for the period of 2007 to 2009. We first mapped and classified timber harvesting as legal and illegal. Next, we identified the municipalities in those States where illegal forest activity is most critical. Later, we overlaid the map of illegal logging on the Protected Areas and land reform settlements so as to identify the areas under the greatest pressure from illegal timber harvesting. Finally, we integrated the information from satellite images with those of the forest control systems in those States.

5.1 Mapping of timber harvesting

We mapped logging using the NDFI images and overlaid that information on the map of forest management plans so as to identify non-authorized logging (illegal and predatory) and authorized logging (forest management). We quantified 543,504 hectares of logged forests in Pará, of which 86% were not authorized and 14% had an authorization for forest management. In Mato Grosso, we mapped 460,134 hectares of logged forests, of which 39% lacked authorization and 61% were authorized (Figure 7).

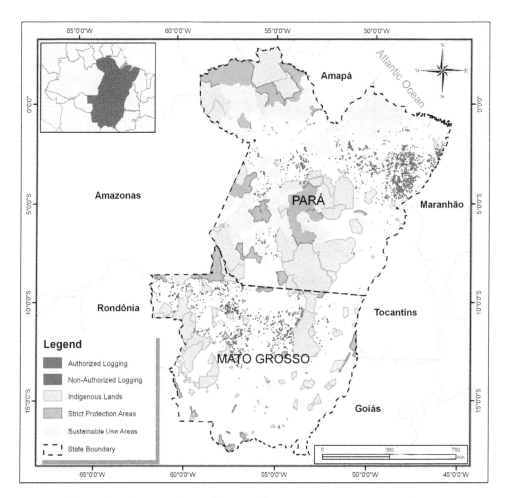

Fig. 7. Authorized and non-authorized logging from 2007 to 2009 in Pará and Mato Grosso states.

5.1.1 Non-authorized logging

Of the 466,979 hectares of forest logged without authorization in Pará between 2007 and 2009, the majority (77%) occurred in 10 municipalities. The remaining 23% were distributed more sparsely among 41 other municipalities. The municipality of Paragominas presented the largest area of non-authorized logging, followed by Rondon do Pará and Goianésia do Pará. In Mato Grosso, there were 179,155 hectares of forests logged without authorization, of which the majority (62%) occurred in 10 municipalities.

The remaining 38% were distributed more sparsely among 32 other municipalities. The municipality of Marcelândia presented the largest area of non-authorized logging, followed by Nova Maringá and Aripuanã.

From 2007 to 2009, illegal timber harvesting in Pará affected 54,874 hectares of forests in Protected Areas. Of that total, 83% was logged in Indigenous Lands (TI) and 17%, in Conservation Units (UC). TI Alto Rio Guamá was the most logged, followed by TI Sarauá and TI Cachoeira Seca. Among the Pará UCs, the National Forests (Flonas) of Jamanxim, Caxiuanã and Trairão are stand out as having the largest volume of timber harvesting. In Mato Grosso, illegal timber harvesting affected 10,524 hectares of forests in Protected Areas: 86% in TIs and 14% in UCs. TI Zoró had the highest amount of logging, followed by TI Aripuanã and TI Irantxe. Among the UCs, an Extractive Reserve (Resex Guariba/Roosevelt) and the Serra de Ricardo Franco State Park (PE) stand out with highest harvest volumes. Monitoring of Protected Areas is extremely important for guaranteeing their integrity and the sustainability of populations that depend on the forest for a living. Thus, environmental agencies can use this tool to restrain devastation of Protected Areas in the Amazon. Additionally, forest concessions in public forest areas such as Flonas need to guarantee income and employment for the population living inside and around those Protected Areas.

In the land reform settlements in Pará, timber harvesting without authorization between 2007 and 2009 affected 53,924 hectares of forests; the majority (75%) in 10 settlements. The most critical situation occurred in the Liberdade Sustainable Development Project (PDS) (50% of the total harvested), followed by the Ouro Branco I and II Collective Settlement Projects (PAC) (12% and 8%). In Mato Grosso, timber logging without authorization in the settlements affected 994 hectares of forests. The most critical situation was the Settlement Project (PA) of Pingos D'água (44% of the total harvested), followed by PA Santo Antonio do Fontoura I (33%). Rural settlement projects in the Amazon hold forest areas with great timber potential. However, in the majority of those projects logging is done in an illegal manner, meaning without a logging license. Programs that encourage forest practices through technical capacity-building for settlers can contribute towards reducing illegal timber harvesting in the settlements and generate income for those families.

5.1.2 Authorized harvesting

For the areas with authorized harvesting, in other words, with forest management, we evaluated the data contained in the Forest Harvesting Authorizations (*Autorizações de Exploração Florestal* - Autef) and in the timber credits issued from 2007 to 2009, in order to verify their conformity or consistency. That information is made available by the State Environmental Secretariats (Sema) in Pará and Mato Grosso, in their systems for forest control, Simlam (Integrated System for Environmental Monitoring and Licensing - *Sistema Integrado de Monitoramento and Licenciamento Ambiental*) and Sisflora (System for Sale and Transport of Forest Products - *Sistema de Comercialização e Transporte de Produtos Florestais*).

In Pará, 277,440 hectares of forests were licensed for management. Of that total, the majority (87%) did not present inconsistencies, while 13% revealed inconsistencies, such as: i) authorization for forest management in area totally or partially without forest cover (6% of cases evaluated); ii) area authorized for management superior to the total area for forest management (4% of cases); and iii) authorization for forest management in area already harvested through logging activities (3% of cases).

In Mato Grosso, 498,783 hectares of forests were approved for forest management, of which the majority (81%) presented no problems and 19% revealed inconsistencies. Those include: i) timber credit commercialized does not correspond to credit authorized (16% of cases); ii) area authorized in deforested area (1% of cases); iii) area authorized greater than the area for forest management (1% of cases); iv) credit issued without authorization for forest harvesting (1% of cases).

Finally, we integrated information from the Autefs with our satellite image base to assess the consistency of forest management performance. In Pará, the largest percentage (45%) of the Autefs evaluated in the satellite image presented no problems, while in 31% it was not possible to make an evaluation because of cloud cover and 24% revealed problems, such as: i) lacking signs of scarring from logging in the images for the period in which the logging authorization was in effect (11% of cases); ii) area of forest management licensed overlaying a Protected Area (5% of cases); iii) logging carried out before issuance of the forest authorization (3% of cases); iv) area licensed for forest management deforested before receiving authorization for harvesting (3% of cases); and v) area logged above the authorized limit (2% of cases).

In Mato Grosso, the same analysis revealed that the majority (78%) presented no problems in the satellite image, whereas 22% revealed problems, which were: i) area was logged above the authorized limit (16% of cases); ii) area licensed for forest management deforested before receiving authorization for harvesting (3% of cases); iii) lacking signs of scarring from logging in the images for the period in which the logging authorization was in effect (1% of cases); iv) plan overlaying a Protected Area (1% of cases); v) logging carried out before issuance of the forest authorization (1% of cases).

The method proposed in this study is capable of monitoring the performance of forest management by timber cutters and forest management licensing by the environmental agencies. This makes it possible to reduce the errors and frauds in the forest control systems during the forest management licensing process and during commercialization of timber.

6. Conclusion

Characterizing the impacts of timber harvesting in the field is essential for determining changes in the structure and composition of the forest submitted to different levels of forest degradation. However, that activity is extremely expensive and lengthy. The advance in techniques for detecting and mapping timber harvesting and integration of that information with data from the field has made it possible to monitor logging (Monteiro et al., 2011) and quantify the loss of carbon from degraded forest in the Brazilian Amazon (Souza Jr et al., 2009).

On the other hand, there is the challenge of putting into operation a system for monitoring timber harvesting at the scale of the Amazon. Logging in the region is predominantly predatory, which has contributed towards an increase in forest degradation and a reduction of the stocks of individual tree species with timber potential. Currently, the Amazon has two systems for detecting deforestation (clearcutting) the

Program for Monitoring Deforestation in the Amazon (*Programa de Monitoramento do Desflorestamento da Amazônia* - Prodes), developed by the Brazilian Space Research Agency (Inpe), and the Deforestation Alert System (*Sistema de Alerta de Desmatamento* - SAD), developed by the Amazon Institute for People and the Environment (Imazon). The method for monitoring timber harvesting proposed in this study can contribute towards reducing illegal logging and improve the quality of harvesting through forest management in the region. With that, we can reduce emissions of CO_2 (Putz et al., 2008) and guarantee the sustainability of the forest-based economy in the Amazon. That method could also contribute towards improving forest management in the Amazon by making it more efficient and transparent.

7. Acknowledgments

We thank the support of the Gordon & Betty Moore Foundation, US Agency for International Development- USAID, US Forest Service and Fundo Vale. Also we thank the NASA/LBA program for funding the forest transect surveys. We would like to thank Dalton Cardoso and Denis Conrado for conducting the remote sensing data processing and Marcio Sales in data statistical analysis.

8. References

Adams, J. B., Smith, M. O., & Gillespie, A. R. (1993). Imaging spectroscopy: Interpretation based on spectral mixture analysis. In V.M. Pieters, & p. Englert (Eds.), Remote Geochemical Analysis: Elemental and Mineralogical Composition, vol. 7 (pp. 145-166). New York: Cambridge Univ. Press.

Amaral, P., Veríssimo, A., Barreto, P., & Vidal, E. (1998). *Floresta para Sempre: um Manual para Produção de Madeira na Amazônia*. Imazon: Belém (In Portuguese).

Asner, G. P., Keller, M., Pereira, R., & Zweede, J. C. (2002). Remote sensing of selective logging in Amazonia-Assessing limitations based on detailed field observations, Landsat ETM+, and textural analysis. *Remote Sensing of Environment*, Vol. 80 N°. 3, pp. 483-496, ISSN 0034-4257.

Asner, G., Keller, M., & Silva, J. N. M. (2004). Canopy damage and recovery after selective logging in Amazonia: Field and satellite studies. *Ecological Applications*, Vol. 14 N°.4, pp. 280-S298, ISSN 1051-0761.

Asner, G. P., Knapp, D. E., Broadbent, E. N., Oliveira, P. J. C., Keller, M., & Silva, J. N. (2005) Selective logging in the Brazilian Amazon. *Science* Vol. 310 N°.5747, pp. 480-482, ISSN 1095-9203.

Cochrane, M., & Souza Jr., C. (1998). Linear mixture model classification of burned forests in the Eastern Amazon. *International Journal of Remote Sensing*, Vol. 19, N°.17, pp. 3433-3440, ISSN: 0143-1161.

Gerwing, J. J. (2002). Degradation of forests through logging and fire in the eastern Brazilian Amazon. *Forest Ecology and Management*, Vol.157, N°.1, p.131-141, ISSN 0378-1127.

Higuchi, N., & Carvalho Jr., J. A. (1994). Biomassa e conteúdo de carbono de espécies arbóreas da Amazônia, In: *Seminário Emissão x Sequestro de CO₂: uma nova oportunidade de negócios para o Brasil*, Porto Alegre. Anais. Companhia Vale do Rio Doce, Rio de Janeiro, pp. 125-153.

Holdsworth, A. R., & Uhl, C. (1997). Fire in Amazonian selectively logged rain forest and the potential for fire reduction, *Ecological Applications*, Vol.7, pp. 713-725, ISSN 1051-0761.

Hurtt, G., et al. (2003). IKONOS imagery for the Large Scale Biosfphere Atmosphere Experimental in Amazonia (LBA). *Remote Sensing of Environment*, Vol. 88, pp. 111-127, ISSN 0034-4257.

John, J. S., Barreto, P., & Uhl, C. (1996). Logging damage during planned and unplanned logging operations in the eastern Amazon. *Forest Ecology and Management*, Vol. 89, pp. 59-77, ISSN 0378-1127.

Martini, A., Rosa, N., & Uhl, C. (1994). An attempt to predict which Amazonian tree species may be threatened by logging activities. *Environmental Conservation, Lausanne*, Vol. 21, N°. 2, pp. 152-162, ISSN 0376-8929.

Matricardi, E. A. T., Skole, D. L., Chomentowski, M. A., & Cochrane, M. A. (2001). Multi-temporal detection of selective logging in the Amazon using remote sensing. Special Report BSRSI Research Advances-Tropical Forest Information Center, *Michigan State University* N° RA03-01/w, 27 p.

Matricardi, E., Skole, D., Cochrane, M. A., Pedlowski, M & Chomentowski, W. (2007). Multi-temporal assessment of selective logging in the Brazilian Amazon using Landsat data. *International Journal of Remote Sensing* Vol. 28, N°. 1, pp. 63-82, ISSN: 0143-1161.

Monteiro, A., Souza Jr, C., & Barreto, P. (2003). Detection of logging in Amazonian transition forests using spectral mixture models. *International Journal of Remote Sensing*, Vol.24, N°.1, p. 151-159, ISSN: 0143-1161.

Monteiro, A. L., Souza Jr, C. M., Barreto, P. G., Pantoja, F. L., & Gerwing, J. J. (2004). Impactos da exploração madeireira e do fogo em florestas de transição da Amazônia Legal. *Scientia Forestalis*, N°. 65, pp. 001-227, ISSN 1413-9324 (In Portuguese).

Monteiro, A. L. (2005). Avaliação de Indicadores de Manejo Florestal na Amazônia Legal Utilizando Sensoriamento Remoto. *Universidade Federal do Paraná-Curitiba*, Dissertação de Mestrado em Manejo Florestal 105p.

Monteiro, A & Souza Jr., C. (2006). Satellite images for evaluating forest management plans. *State of the Amazon*, N°.9, Belém: Imazon, 4 p.

Monteiro, A. L., Lingnau, C., & Souza Jr., C. (2007). Classificação orientada a objeto para detecção da exploração seletiva de madeira na Amazônia. *Revista Brasileira de Cartografia*, N°. 59/03, Dez. 2007. 10 p, ISSN 1808-0936 (In Portuguese).

Monteiro, A., Brandão Jr., A., Souza Jr., C., Ribeiro, J., Balieiro, C., & Veríssimo, A. (2008). Identificação de áreas para a produção florestal sustentável no noroeste do Mato Grosso. Belém: Imazon, 68 p. (In Portuguese).

Monteiro, A., Cardoso, D., Veríssimo, A., & Souza Jr., C. (2009). Transparency in Forest Management-State of Para 2007/2008. Imazon, Belém.

Monteiro, A., Cruz, D., Cardoso, D. & Souza Jr., C (2011). Avaliação de Planos de Manejo Florestal na Amazônia através de imagens de satélites Landsat, Anais XV Simpósio Brasileiro de Sensoriamento Remoto-SBSR, Curitiba-PR, Brasil, INPE p. 5615, ISBN 978-85-17-00057-7.

Nelson, B. W., Mesquita, R., Pereira, J. L. G., Souza, S. G., Batista, G. T., Couto, L. B. (1999). Allometric regression for improved estimate of secondary forest biomass in the central Amazon. *Forest Ecology and Management*, Vol.117, p. 149-167, ISSN 0378-1127.

Overman, J. P. M., Witte, H. J. L., & Saldarriaga, J. G. (1994). Evaluation of regression models for aboveground biomass determination in Amazon rain forest. *Journal Tropical Ecology* Vol. 10, N°. 2, pp. 207-218, ISSN 0266-4674.

Pereira Jr, R., Zweed, J., Asner, G., & Keller, M. (2002). Forest canopy damage and recovery in reduced-impact and conventional selective logging in eastern Para, Brazil. *Forest Ecology and Management*, Vol. 168 (1-3), pp.77-89, ISSN 0378-1127.

Pereira, D., Santos, D., Vedoveto, M., Guimarães, J & Veríssimo, A. (2010). *Fatos Florestais da Amazônia 2010*. Imazon Belém, 126p. (In Portuguese).

Putz, F. E., et al. (2008). Improved Tropical Forest Management for Carbon Retention. *PloS Biology*, pp. 1368-1369, N°. 7, ISSN 1545-7885.

R Development Core Team (2010). R: A language and environment for statistical computing. R Foundation for Statistical Computing,Vienna, Austria. ISBN 3-900051-07-0, URL <http://www.R-project.org/>

Read, J. M., Clark, D. B., Venticique, E. M & Moreira, M. P. (2003). Applications of merged 1-m and 4-m resolution satellite data to research and management in tropical forests. *Journal of Applied Ecology* Vol. 40, N°. 3, pp. 592-600, ISBN 1365-2664.

Souza Jr, C. & Barreto, P. (2000). An alternative approach for detecting and monitoring selectively logged forests in the Amazon. *International Journal of Remote Sensing*, 21, p. 173-179, ISSN: 0143-1161.

Souza Jr, C., Firestone, L., Moreira, L., Roberts, D. (2003). Mapping Forest degradation in the eastern Amazon from SPOT 4 throgh spectral mixture models. *Remote Sensing of Environment*, 87, p.494-506, ISSN 0034-4257.

Souza Jr, C., Roberts, D. A., & Cochrane, M. A. (2005). Combining spectral and spatial information to map canopy damage from selective logging and forest fires. *Remote Sensing of Environment*. Vol. 98, pp. 329-343, ISSN 0034-4257.

Souza Jr., C., & Roberts, D. (2005). Mapping forest degradation in the Amazon region with Ikonos images. *International Journal of Remote Sensing*, Vol. 26, N°. 3, pp. 425-429, ISSN: 0143-1161.

Souza Jr., C. M., Cochrane, M. A., Sales, M. H., Monteiro, A. L., & Mollicone, D. (2009). Integranting forest transects and remote sensing data to quantify carbon loss due to Forest degradation in the brazilian amazon. *Forest Resources Assessment Working paper* 161, FAO, 23 p.

Stone, T. A. & Lefebvre, P. (1998). Using multi-temporal satellite data to evaluate selective logging in Para, Brazil. *International Journal of Remote Sensing*, Vol. 19, pp.2517-2626, ISSN: 0143-1161.

Veríssimo, A., Barreto, P., Mattos, M., Tarifa, R., & Uhl, C. (1992). Logging impacts and prospects for sustainable forest management in an old Amazonian frontier- the case of Paragominas. *Forest Ecology and Management*, Vol. 55 N°. 1-4, pp. 169-199, ISSN 0378-1127.

4

Sustainable Management of Lenga (*Nothofagus pumilio*) Forests Through Group Selection System

Pablo M. López Bernal[1,2,3], Guillermo E. Defossé[1,2,3],
Pamela C. Quinteros[1,2] and José O. Bava[2,3]
[1]*Consejo Nacional de Investigaciones Científicas y Técnicas – CONICET*
[2]*Centro de Investigación y Extensión Forestal Andino Patagónico – CIEFAP*
[3]*Universidad Nacional de la Patagonia San Juan Bosco – UNPSJB*
Argentina

1. Introduction

1.1 Distribution and environmental gradients of lenga forests

Lenga (*Nothofagus pumilio* (Poepp. *et* Endl.) *Krasser*) is a native tree species widely distributed in the Andean forests of Patagonia. In Argentina, lenga forests cover almost the entire length of the sub-Antarctic forests on the eastern slopes of Andean Cordillera, from the 35 ° 35 'S latitude parallel in the province of Neuquén to the 55 ° S in the province of Tierra del Fuego (Figure 1). This species usually occupies the upper portion of the altitudinal limit of woody vegetation (up to 2000 masl) in its northern distribution area, while it grows near sea level in its southern distribution area in Tierra del Fuego (Donoso Z., 1995, Tortorelli, 2009, Veblen et al., 1977).

Lenga is adapted to grow under a great variety of soils, environmental conditions, and disturbance regimes (Schlatter, 1994). In fact, this species could be found in areas in which average annual precipitation may reach 500 mm year^{-1} (under a Mediterranean type of climate), to others reaching 3,000 mm year^{-1} (under either iso-hygro or Mediterranean type of climates). Lenga is also capable of supporting extreme temperatures, from mean annuals of 3.5 to 4 °C in upper altitudinal areas (Schlatter, 1994) to 7 to 9 °C in milder areas at lower altitudes or also near sea level.

In the northern part of its distribution, lenga grows under a typical Mediterranean climate, with precipitation concentrated during winter and early spring as either rain or snow, followed by a dry and mild period during summer and early fall. Going south, this regime gradually changes to more iso-hydric conditions, being precipitation more evenly distributed along the year. In the northern part of its distribution and up to the 52 ° S, however, the amount of annual precipitation is greatly influenced by the barrier that imposes the Andean Cordillera, which creates one of the most spectacular precipitation gradients of the world. There, the western humid winds coming from the South Pacific Ocean discharge most of the precipitation as they go upward to the upper parts of the Andes, passing to the eastern slopes as more dry air masses that rapidly lose their humidity content. This makes that upper mountain ranges near the border with Chile may receive 5000 mm of precipitation per year, while in less than 50 to 80 km toward the Patagonian steppe, precipitation sharply diminishes to ca 500 mm annually (Barros et al., 1983, Jobbágy et al., 1995, Veblen et al., 1977). To the South of the 52 and up to the 55 ° S parallel in the island of Tierra del Fuego, a regular rainfall pattern occurs, with rainfall evenly distributed throughout the year (Burgos, 1985);

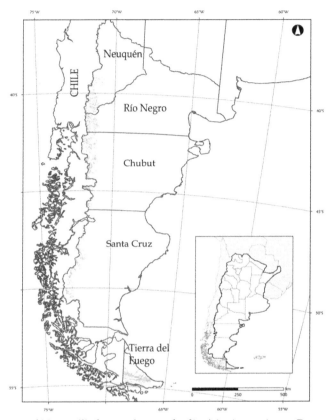

Fig. 1. Distribution of *N. pumilio* forests (green shading) in Argentinean Patagonia.

1.2 Disturbance regime

Throughout its wide distribution area, lenga stands are clearly distinguishable from other component of the Andean forests, being composed of simple monospecific structures with narrow ecotones (Donoso Z., 1995). However, given the different environments in which it develops, lenga presents different structures and regenerative dynamics, mainly associated with the frequency, magnitude and severity of disturbances such as windstorms, fires, avalanches, landslides, or the falling of senescent trees (Donoso Z., 1995, Veblen et al., 1996). As a consequence of these disturbances, lenga stands may present either even or uneven-aged structures, both situations representing extremes in a range of different possible structures. At the southern end of lenga distribution in Tierra del Fuego, tree falls usually occur due to wind storms, and this result in even-aged young structures (Rebertus et al., 1997). Furthermore, the same wind storms that cause large falls may also lead to formation of small gaps in the forest canopy. Uneven-aged structures are usually originated in mature forests located in favorable sites at low altitudes having low frequencies of catastrophic disturbances or human interventions. In these areas, the falling out of senescent trees may promote the opening of gaps or patches of about 0.1 ha (Bava, 1999, Veblen & Donoso Z., 1987) in which regeneration begins. These patches generally possess favorable undergrowth conditions which allow the formation of small clumps of saplings. The result of this process is a multi-aged and multi-strata forest, even when the formation of these gaps may be an episodic phenomenon (Rebertus & Veblen, 1993).

In relation to its tolerance to shadow, lenga has been classified as either "purely heliophilous" (Mutarelli & Orfila, 1971), "semi-heliophilous" (Tortorelli, 2009), "medium tolerant" (Rusch, 1992) to "semi-tolerant" (Donoso Z., 1987). These controversial or even opposite classifications are probably due to the different habitats these descriptions came from. It has been well established that for many species, proper development may depend on the limiting resources a given environment may have (Choler et al., 2001) so the radiating needs of lenga regeneration may significantly vary depending on a set of other environmental factors (Rusch, 1992).

In sites with high rainfall levels (i.e. South of Tierra del Fuego and West of Chubut province), lenga regeneration is established even after major disturbances affecting up to hundreds or even thousands of hectares (Rebertus et al., 1997, Veblen et al., 1996). The same occurs in forests affected by intensive forest harvesting (Gea Izquierdo et al., 2004, Mutarelli & Orfila, 1971, Rebertus & Veblen, 1993, Rosenfeld et al., 2006). By contrast, in sites with water deficit during the summer, as in the northern sector of lenga distribution in Río Negro, Neuquén and Chubut provinces, regeneration cannot be established with low canopy coverage (Bava & Puig, 1992). In these areas, regeneration establishment is strongly influenced by water availability and usually occurs in small gaps caused by falling trees (Rusch, 1992).

Recent studies have analyzed the effects of micro-environmental factors on the establishment and growth of lenga seedlings in natural gaps. These studies showed that in the driest sites of lenga distribution, the shade generated by individuals from the edge of the gaps and the presence of coarse woody debris, produce a facilitator effect on seedling establishment (Heinemann & Kitzberger, 2006, Heinemann et al., 2000). Seedling survival in these xeric sites have been positively related to water availability, while in mesic sites survival seems to be controlled by both water availability and light (Heinemann & Kitzberger, 2006, Heinemann et al., 2000).

2. History of productive use of *N. pumilio* forests in Argentinean Patagonia

Most of lenga productive forests in Argentinean Patagonia, owned by either private or state sectors, started to be exploited at the beginning of the XXth century, but did not reach significant levels of harvest until mid-century, with the emergence of large sawmills that used almost exclusively high quality timber. Since then, the closing down of these large sawmills and the gradual installation of small and medium sawmills generated changes in harvesting techniques, extraction rates, and final products. These changes were usually marked by the lack of effective control policies by the state administration, which lead to the absence of sustainable management practices. Furthermore and to worsen this situation, these forests have been traditionally used as summer pastures for cattle, which in many cases has caused the degradation of the understory, with long delays in, and even preclusion of, regenerative processes.

The objective of this chapter was to analyze the evolution of productive schemes of lenga forests along their history of use, which will help us understand the overlap of strains on this resource, impacts on their conservation status, and the difficulties that currently have the implementation of sustainable management systems. For this purpose, we got information derived from published analyses, statistical records of the Forest Administration, analyses of historical harvesting, and of the impacts of livestock on forest regeneration. This information is presented for two contrasting situations, located one in the northern lenga distribution area in Chubut province and the other in its southern distribution area in Tierra del Fuego.

2.1 Pre-industrial

The original inhabitants of continental Patagonia were mostly nomadic Indian tribes that depended largely on the guanaco (*Lama guanicoe Muller*) for their livelihood. These tribes used ecotone and steppe areas of and did not settled in the Andean forests, although there are some examples of communities who lived associated with *Araucaria araucana (Molina) K. Koch* forests in northern Patagonia. Lenga forests, located at higher altitudes, were only occasionally used as firewood in the journeys crossing the Andes (Musters Chaworth, 1871). In Tierra del Fuego, unlike continental Patagonia, guanaco used lenga forests as part of its habitat, perhaps due to the absence of its natural predator, the puma (*Puma concolor* Linnaeus). Some Indian tribes lived much of the year in these interior forests, while others were established on the shores of the Beagle Channel, all surrounded by lenga forests. In this region, the use of lenga for small constructions and canoes, although in small scale, has been reported (Bridges, 2000). The major effect of indigenous peoples on Patagonian forests has been the recurrent employment of fire, either for hunting purposes or used as a communications signal (Kitzberger & Veblen, 1999).

During white settlement, cracked poles, rustic tables, or shingles, were widely used products from lenga forests, but undoubtedly, fire was the most devastating factor affecting them. In the Argentine sector of Tierra del Fuego, an estimated 20,000 ha were burnt in the early twentieth century. Contemporarily and in an attempt to open land for sheep raising, pioneers in the Chilean Patagonia initiated what could now be called catastrophic fires, burning large portions of lenga woodlands (2.8 million ha, Fajardo & McIntire, 2010), reducing their original area by a half (Otero Durán, 2006). In the rest of its distribution area, thousands of hectares were also burnt, although reliable data are not available (Willis, 1914). The recovering of lenga forests after those fires depended on a multiplicity of factors, among which the availability of safe sites (*sensu* Harper, 1977) for seed germination and seedling establishment, and the grazing pressure exerted on the burned sites played a crucial role. The outcomes in former lenga forests were then open fields to raise sheep or the slow recovery of lenga forests. After that beginning and in the mid 40´s, factors such as the strengthening of national protected areas, the decline of sheep production and the displacement of rural populations modified this process of impoverishment or forest clearance, at least at regional level. Though, the lenga forests that were formerly used as summer ranges for sheep gradually changed to cattle grazing areas. It is interesting to note that the introduction of cattle ranching in the area has a vague origin, as the early explorers (Musters Chaworth, 1871), cite the existence of wild cattle in the forests of the region already in 1870, possibly coming from Valdivia, Chile (settled around 1600), or escaped from the cattle drives that native communities transported from Argentina to Chile.

2.2 The beginnings of industrial use: High grading

The first forest industries installed in the early twentieth century, either for medium or small sawmills, or for wood veneer production, were characterized by softly logging, cutting only healthy, medium-sized trees. Stem rots, caused mostly by fungi of the genera *Postia* and *Piptoporus*, is a very common phenomenon that affects lenga trees, being very important in old age trees. This determined that in virgin forests, only a small proportion of trees contained good quality timber. For that reason and in general, forest workers used to cut down only trees in good health status, medium-sized (40 - 50 cm DBH), which generally did not exceed 10% of forest trees in a stand. This type of "soft logging" was locally known as "floreo" (high grading).

2.3 Forest management plans

The first Argentine Forest Law was put in force in 1948. While the concept of forest management, as a synonym of timber production, was prevalent in that law, it included articles about protection of soil, water and biodiversity. Although it mandated for the implementation of Forest management plans, its principles and regulations were applied sparingly. As a result, logging continued in public forests in an unplanned way. In the mid 50´s of the XXth century, the first forest management plans were designed and applied by Croat forest engineers, who arrived to Argentina after the World War II. These plans represented a breakthrough for the understanding of lenga forests, but had little practical effects on forest management due to the weaknesses of the Administrative forestry services of the Patagonian provinces.

By the 80's, the practice of giving access to cut lumber in public or private forests depended on the approval of forest management plans by the provincial forest service, practice that became usual. However, these were just cutting plans, without long term planning horizon and being controlled at different levels of implementation.

Silvicultural aspects were changing over time with the evolution of knowledge about forests dynamics, from the early experiences on shelterwood systems in Chile (Cruz M. & Schmidt, 2007), clear-cut in Argentina (Mutarelli & Orfila, 1971), to the currently used alternatives, ranging from a group selection system (Bava & López Bernal, 2005) up to a variation of shelterwood systems with dispersal and-or aggregate retention (Martínez Pastur et al., 2009).

2.4 Cattle

As already mentioned, the first activity developed in Patagonian was sheep ranching. Near the Andes, the usual ranching scheme was a system that alternated winter grazing (locally called invernadas) in low areas with summer grazing areas at higher altitudes in the forest (called veranadas). There are plenty of examples of this system in mountain areas around the world. In the mid-twentieth century, with increasing population established in the area, cattle raising was becoming important, with the same production scheme.

In lenga forest ecosystems of Patagonia, herbivory causes severe impacts, because this species is palatable to both wild (camelidae, deer and leporidae), and domestic livestock, and heavy grazing can prevent forest regeneration (Veblen et al., 1996). In Argentina, lenga forests suitable for timber production are mainly concentrated in the provinces of Chubut and Tierra del Fuego. Lenga forests in the province of Chubut are also a very important part of traditional cattle management, which similarly to what formerly occurred with sheep, alternates winter fields in the steppe with summer fields at the mountains (York et al., 2004).

Therefore, 19 % of the forests suitable for timber production are potentially degraded by cattle grazing (Bava et al., 2006). In Tierra del Fuego, by contrast, is the wild guanaco (*L. guanicoe*) which has a negative impact on regeneration. This impact has been reported especially in forests located northwards of Fagnano Lake and in the Chilean side of the Isla Grande de Tierra del Fuego (Cavieres & Fajardo, 2005).

It is possible to distinguish between direct and indirect effects on forest regeneration caused by large herbivores. The direct consumption of seedlings keeps the lenga regeneration stunted (Perera et al., 2004) and multi-stemmed (Bava & Rechene, 2004). The consumption of other species may cause a decrease in the diversity of the understory. The transport of seeds through feces allows the introduction of exotic species (Bava & Puig, 1992). Immersed in lenga forests is frequent to find meadows, locally called "mallines" in the province of Chubut or "vegas" in Tierra del Fuego. These meadows are highly valued by farmers because of their high productivity in forage species. Due to the existence of meadows near the forests, and the little vegetation cover that characterizes the undergrowth of lenga (Lencinas et al., 2008), an intensive use of resources and a great impact on regeneration and understory forest areas close to the meadows has been reported (Quinteros, unpublished data). The changes that livestock generated in the understory, such as increased coverage of grazing tolerant species, mainly exotic, constitutes an indirect effect on the development of lenga regeneration, because the high grass coverage competes for water resources with lenga seedlings, affecting their growth and development (Quinteros, unpublished data).

3. Group selection system in *N. pumilio* forests

3.1 General concepts

The selective silvicultural systems are characterized by generating uneven-aged stands where regeneration layer strongly interacts with the mature forest, and this interaction could either be favorable or unfavorable for seedlings or saplings of different species (Daniel et al., 1979). With this method, individual trees are removed (or small groups of them), opening small gaps that can be used by tolerant species. Harvesting procedures require frequent partial cuts, where the harvest interval is called "Cutting Cycle" and there is not a rotation age where all production is harvested, as in the even-aged methods (Daniel et al., 1979).

In the Group Selection System (GSS), harvested trees are pooled in small groups (typically up to 10 mature trees), thus creating gaps in the canopy larger than the individual selection cuttings, which are better suited to the requirements of semi-tolerant species, as is the case of lenga (Bava & Rechene, 2004). It also provides some advantages of even-aged stands, as the saplings grow in conditions of intra-cohort competition. This competition favors the production of better shaped stems, while the harvest is partially concentrated, reducing its costs and minimizing the damages from falling trees (Daniel et al., 1979).

Under this scheme, some decisions that have to be taken are (Davis & Johnson, 1987):

• Cutting cycle: time between harvest entries on each stand.
• Reserve growing stock level: residual volume or basal area (BA) immediately after harvest.
• Group, patch or gap size: defined in function of objective species requirements.

The historical origin of this type of management helps gauge its applicability and scope in different productive forest systems. Uneven-aged forest management had its origins in Central Europe, where since the twelfth century exist harvest protocols regulating the forest extraction, by limiting the numbers of individuals or the volume to be harvested (Becking, 1995). However, the real practice was a selective extraction of the best stems (high-grading) without control policies that would ensure the regeneration and future productive potential of the forest, affecting negatively the productive quality of large areas. For this reason, between the fifteenth and eighteenth centuries this type of logging was progressively sidelined (Puettman et al., 2009), and new forest practices, such as clear-cutting, emerged (Becking, 1995). In the early nineteenth century, an alternative selection system was formalized for various regions of Europe, where clear-cuts were banned and where the landowners of small forest stands were especially interested in the high frequency of harvesting (short cutting cycles) thus maintaining a continuous cash flow (Puettman et al., 2009).

As mentioned in the previous section, the historical context of lenga forests in Patagonia, and particularly in the north of its distribution, is in some ways comparable to the origins of the implementation of selective cuts. The predominance of small and medium producers, the low productivity of these forests and the low control capacity of state agencies has meant that in most cases the harvest have been a high-grading, often of low intensity. Thus the GSS, where harvesting is simultaneous with other silvicultural tasks (such as thinning or regeneration release) on the one hand represents an economically feasible objective for local producers, and on the other a simplification of the control tasks posed by the state.

Moreover, the prolonged rotation periods needed for lenga forests and the brief history of the implementation of these schemes in Patagonia prevents direct observation of long-term management examples. Given this situation, models of conservation and sustainable management based on emulation of natural disturbance regimes are very attractive for developing sustainable management practices (Perera et al., 2004).

From this view, various management systems have been proposed through intense felling as clear-cuts or shelter-wood cuttings (Arce et al., 1998, Martínez Pastur et al., 2009); with the intention of imitating mass disturbances that naturally occur, especially in southern Patagonia. However in Chubut province, the rainfall regime with wet winters and prolonged dry periods in summer, prevents the proper regeneration establishment in large areas subject to direct sunlight (Rusch, 1992). For this reason, an adaptation of Group Selection System (1997) is currently proposed for these sites, imitating the predominant disturbance of gap dynamics (Bava, 1999, Veblen & Donoso, 1987, Veblen et al., 1981, Veblen et al., 1980). This promotes the establishment of regeneration patches formed by felling from one to six trees (Antequera et al., 1999, López Bernal et al., 2003).

3.2 Adaptation of GSS to *N. pumilio* forests
3.2.1 Definition of canopy gap

The minimum unit for the application of different treatments in a group selection system is the "forest patch" or "canopy gap". However, the definition of canopy gap or its size is often unclear. On one hand, there are different definitions of the "canopy gap limit" and on the other, there are various ways to simplify its form. Additionally, there are several field methods for gap size measurement. López Bernal et al. (2010) compared different ways of this three issues, specifically:

i.   Gap limit definitions: there are two main schools of thought defining this parameter. One is proposed by Brokaw (1982), who defined the gap as a "hole" in the forest that extends across all levels to an average height of two meters above the ground, and whose boundaries are defined as vertical walls. The space calculated by this method is usually called the "canopy gap" (Figure 2). However, this method has been criticized because it underestimates the area affected by the gap (Popma et al., 1988). The other definition was proposed by Runkle (1981), based on the concept of an "expanded gap" whose limits extend to the base of the bordering trees. Runkle argued that this method has the advantage of including the area where light availability is directly and indirectly influenced by the gap.

ii.   Calculation methods: regardless of the gap type (i.e. the definition of its limits), there are several methods to calculate or estimate the surface area of a gap. These methods mainly differ in the degree of form simplification, i.e. how well they capture boundary irregularities, moving from ellipses to polygons, octagons or hexadecagons, either with straight sides or with sections of an ellipse (Brokaw, 1982, Green, 1996, Lima, 2005, Runkle, 1981, Zhu et al., 2009).

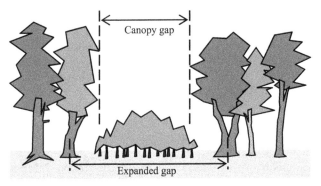

Fig. 2. Canopy and expanded gap scheme.

iii. Field Methods: Finally, different methods, such as measuring directions and distances from gap center or the triangles method (Lima, 2005), may be applied to measure the variables needed to calculate gap size. These methods may be more or less effective depending on the characteristics (such as understory density and height) of the forest being studied. The optimal field method must also be evaluated in terms of its simplicity of operation, time requirement, necessary tools, etc.

The three issues listed above are all based on the conception of a gap as a surface. The relationship between gap diameter and canopy height has also been used as a reference parameter in some studies (Albanesi et al., 2008, Minckler & Woerheide, 1965, Runkle, 1985), especially where gap creation has been used as a management activity. Canopy height is a parameter with a direct influence on the amount of received radiation. Therefore, the addition of canopy height in any calculation may lead to a significant improvement in the accuracy of gap size estimation.

López Bernal et al. (2010) concluded that the Polygonal Expanded Gap Diameter / dominant canopy Height ratio (from now on D/H) is an expeditious method to characterize gap size. This method not only allows the estimation of the incident radiation, but also the comparison of gaps of different stands and even of different species (Albanesi et al., 2008, OMNR, 2004). The method also incorporates dominant canopy height, which improves gap characterization at different sites. A range for this variable for lenga forests is between 14 and 30 m, and makes D/H an adaptable parameter. The strong correlation between D/H and incident radiation makes this parameter a good radiation predictor for gaps in a broad range of gap sizes and with canopies of different heights, and represents a useful tool, both to define silvicultural guidelines and to carry out forest ecological studies.

3.2.2 Tree marking guidelines

The main strategy proposed by Bava and Lopez Bernal (2005, 2006) for marking in virgin or high graded forests, focuses on trees from which it is currently possible to gain good quality logs, or on young healthy trees showing high timber potential. If these trees exceeded the minimum diameter at the breast height (DBH) of 35 cm (or 40 cm if they had smooth bark), they are felled in order to open or expand the gap, but if they not exceed that diameter, their growth is favored by cutting or girdling competitor trees. Thus, the procedure identifies almost homogeneous small patches within the stand, and depending on their structure, the decision between this three alternatives is made:

Gap opening: Operation for the opening of a gap by felling healthy mature trees and girdling old rotten ones, where a small regeneration patch must grow successfully.

Gap release: Operation oriented to release seedlings patches in old gaps by cutting old over-matures neighbor trees.

Thinning: Release of young healthy saplings or poles (15-30 cm DBH), by cutting competitors trees, mostly from the same cohort.

These general rules can become more specific, taking into account the rainfall level (Figure 3):

Mesic sites (Rainfall > 1100 mm/year)

Stage 0: Patch of mature trees.

i.   In a patch composed of mature trees, new gaps with D/H between 1.5 -2 have to be open. We expect that after a rotation of 35 years, the regeneration here will be at least 5 m height.

Stage 1: Gap with seedlings less than 1 m height.

ii. If the available light is not enough for a successful growth, we enlarge the gap up to H/D 2. That allows maximizing the height growth up to 25-30 cm/year.

Stage 2: Gap with seedlings higher than 1 m.

iii. In this case, it is possible that the regeneration losses his form and vigor because of inappropriate light conditions, and the opening of a new gap as in situation 0 is needed. In the other case, with the regeneration in good condition, we enlarge the gap up to D/H = 2. In that case we expected height growth rates similar to that mentioned in situation 1.

iv. If the gap is colonized by seedlings with good growth, there is the possibility that some dominant seedlings with bad form are preventing the proper development of the rest; in this case they should be removed by cutting or girdling.

Stage 3: gap with saplings.

v.   In this situation the gaps limits are unclear and it is possible to recognize the good quality saplings or poles with the potential to reach commercial sizes (DBH > 40 cm). They have to be released from their main competitors.

Xeric sites (Rainfall < 1100 mm/year)

Stage 0: Patch of mature trees.

vi. In this situation, where individuals from the canopy have reached an age and size that made them suitable for harvesting, new gaps should be open with D / H between 0.8

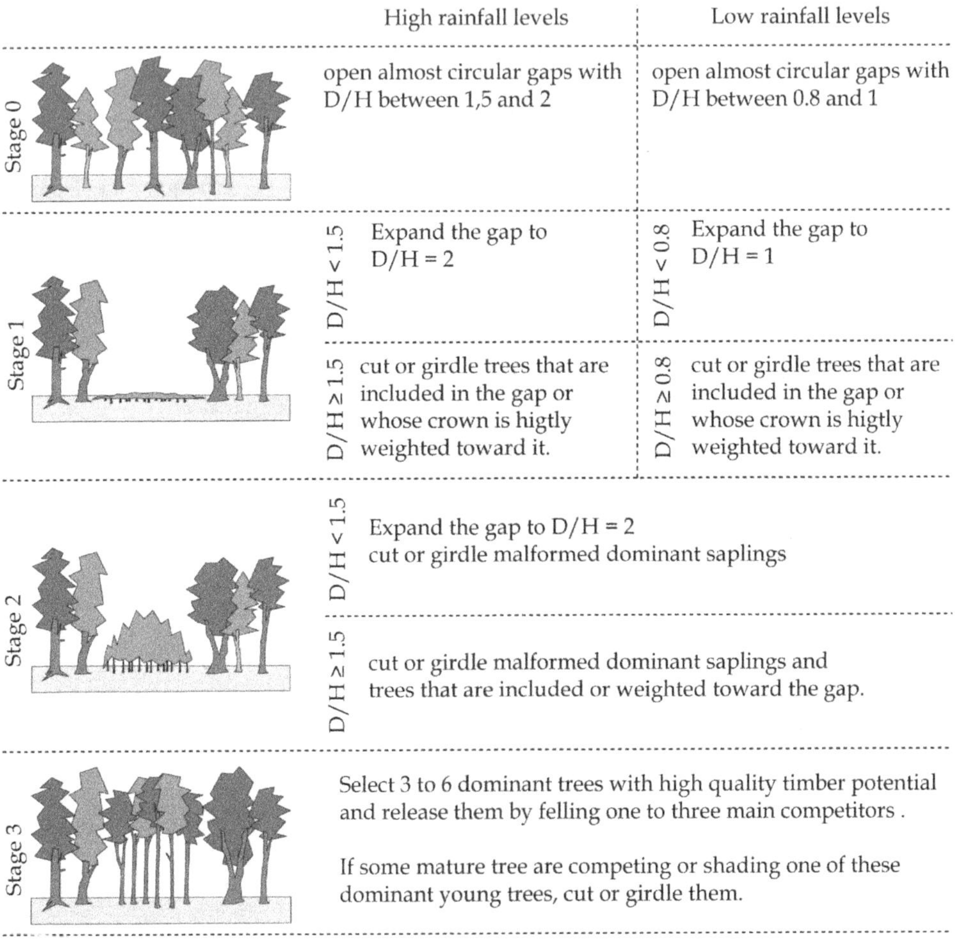

	High rainfall levels	Low rainfall levels
Stage 0	open almost circular gaps with D/H between 1,5 and 2	open almost circular gaps with D/H between 0.8 and 1
Stage 1	D/H < 1.5 — Expand the gap to D/H = 2 D/H ≥ 1.5 — cut or girdle trees that are included in the gap or whose crown is higtly weighted toward it.	D/H < 0.8 — Expand the gap to D/H = 1 D/H ≥ 0.8 — cut or girdle trees that are included in the gap or whose crown is higtly weighted toward it.
Stage 2	D/H < 1.5 — Expand the gap to D/H = 2, cut or girdle malformed dominant saplings D/H ≥ 1.5 — cut or girdle malformed dominant saplings and trees that are included or weighted toward the gap.	
Stage 3	Select 3 to 6 dominant trees with high quality timber potential and release them by felling one to three main competitors. If some mature tree are competing or shading one of these dominant young trees, cut or girdle them.	

Fig. 3. Schematic representation of the marking procedure.

and 1. Situations where seedlings are already present should be preferred. Thus it is expected that after a short cycle of about 35 years regeneration has reached a height of about 3 m.

Stage 1: gap with seedlings less than 1 m height.

vii. Given this situation, it is recommended to take no action unless you notice the presence of an isolated adult tree stocked in the gap (not as border tree), which should be girdled.

Stage 2: gap with seedlings higher than 1 m.

viii. If the gap has a size from D/H < 1, it should be expanded to reach D/H = 1.5 to 2. The height growth in this condition will reach 15-20 cm/year.

ix. If the gap is colonized by seedlings with good growth, it is the possible that dominant seedlings with bad form are preventing the proper development of the rest; in this case they should be girdled.

Stage 3: gap with saplings.

x. In these cases the gap limits are unclear and individuals have reached a size that allows us to identify those with potential to reach the appropriate size for harvesting (e.g. DBH greater than or equal to 40 cm). They should be released from its major competitors to maximize their growth.

There are three key issues for the success of group selection system in lenga forests. First, regeneration must be installed and growing properly in the gap as to reach their final height with a good stem form. Second, the lateral crown growth of trees bordering the gap must not interfere with the proper development of saplings. Finally, the remaining volume stock after each intervention must maintain its stability until the next harvest. Here we review these three issues, focusing on the aspects that should be taken into account in the definition of guidelines for forest management of lenga by group selection system.

3.3 Regeneration requirements
3.3.1 Regeneration establishment

Several studies in the northern area of distribution of lenga (xeric sites) have concluded that the establishment of the regeneration of this species is strongly dependent on the availability of water during the growing season. In the drier sites located to the east, regeneration is only installed on microsites that, because of being shadier or because of the protection of course woody debris, remain wetter in summer (Heinemann & Kitzberger, 2006, Heinemann et al., 2000, Rusch, 1992). Thus, the position within the gap is a decisive factor for the recruitment of regeneration in drier sites, whereas in moist sites the position does not influence seedling survival. The same authors found that the initial growth of seedlings in the driest sites was greater in the shady parts of the gap, while in wetter sites the initial growth was higher in the center, concluding that moisture and light availability are the limiting factors for recruitment and early growth for sites with lower and higher levels of precipitation, respectively.

Thus, on the sites without drought stress during the summer, as in the western sector of the distribution of lenga in the province of Chubut, larger gaps will be more adequate, in which the interaction between the canopy and the regeneration is lower. By contrast, in sites with a high hydric deficit during the growing season, located east of the distribution of this species, the facilitating effects in microsites protected from direct sunlight and with lower evapotranspiration by the canopy, outweigh the effect of competition for other resources. As a result, smaller gaps will present the highest values of recruitment.

3.3.2 Saplings growth

Having established the regeneration, the requirements for their development change as the seedlings grow in height and their roots explore the soil profile (Callaway & Pugnaire, 2007). Figure 4 shows the values of mean annual increments in height (MAIh) for every level of precipitation and gap size. These values were estimated by a mixed ANCOVA model, in which sapling height was included as a co-variable (López Bernal, unpublished data). During the first 20 years since the gap opening, in the sites with higher levels of precipitation, the dominant seedlings located in the central sector of the gap showed higher growth in larger gaps ($p = 0.03$ and $p = 0.045$ for 0-10 and 10-20 years respectively). By contrast, in sites with lower average annual rainfall, there is a tendency for smaller gaps to show higher height growth, especially during the first 10 years.

Summarizing, we can infer that during the first 20 years since the opening of the gaps, the growth of regeneration is determined by light availability in moist sites and water availability in dry sites, with average values of about 22 cm/year and 15 cm/year, respectively, showing a decrease in the differences due to rainfall with the gap age.

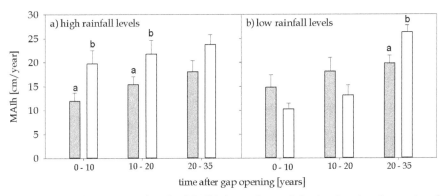

Fig. 4. Mean annual increase in height (MAIh) for each precipitation level and gap size class (small gaps = gray bars, large gaps = white bars) along a 35 years cutting cycle. Different symbols indicates significantly different means (Fisher's posthoc test, α = 0.05).

Moreover, the growth data for gaps between 20 and 35 years old shows that at this stage the saplings grew independently of the availability of water, at least enough to keep differences between the sites with higher and lower levels of precipitation. These observations are consistent with several studies which reported that the balance between facilitation and competition interactions usually tends toward negative values when the "facilitated" individual, approaches the age of maturity (for a comprehensive review of this phenomenon see Callaway & Pugnaire, 2007, pp 240).

3.3.3 Saplings density
Density of seedlings in gaps is often highly variable. During the first years after the creation of gaps, density is strongly determined by the availability of water in the soil, so in places with water deficit during the summer, a greater density is usually observed in the shady gap borders or in microsites caused by the presence of coarse woody debris (Heinemann & Kitzberger, 2006). However, with the subsequent development of the seedlings and the processes of mortality, linked to competence or because of the small disturbances that occur within the gaps (such as total or partial collapse of one of the trees limit), these patterns are lost. For example, it has been observed that in gaps between 20 and 35 years old, significant differences in saplings density between different parts of the gap are not detected (Figure 5, Lopez Bernal et al. Unpublished data). On the other hand, considering only the central part of the gap, there is also great variability, which prevents detect possible influences of gap size or rainfall levels.

3.4 Lateral crown growth of trees bordering the gap
The average closing rate of gaps due to lateral growth of bordering trees is approx. 19 cm/year. This is high enough so that can occur the gap healing before that regeneration can reach the upper stratum (López Bernal et al. unpublished data).

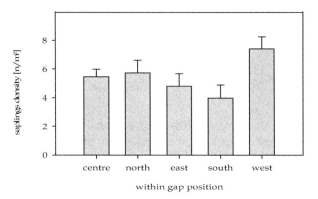

Fig. 5. Average seedling density at different locations within 20-35 years old gaps.

Figure 6 represents the two mechanisms of gap healing (i.e. lateral crown growth of bordering trees and regeneration height growth), indicating the time needed for them to close gaps of different sizes (ordinates). In general, larger gaps require more time for healing by crowns growth and less time for healing by regeneration growth. Thus, the curves representing each mechanism are cut at the point corresponding to the gap size that allows the regeneration to reach the canopy just before the crown growth of bordering trees prevents it. It can also be inferred how long will it take for this to happen (abscissa).

Thus, the ① arrow represents the development of a gap in a humid stand, where it is feasible to open a gap with D/H between 1.5 and 2, favoring the seedlings installation and saplings development until its final height. Moreover, the ② and ③ arrows represent the development of a gap in a xeric stand, where it is necessary to open smaller gaps to ensure seedling establishment, but after a 35 years cutting cycle is necessary to enlarge the gap to prevent the healing by the lateral crown growth of bordering trees.

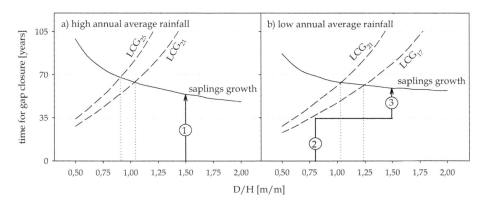

Fig. 6. Necessary time to close gaps of different sizes (D/H) through the height growth of regeneration (solid lines) or the lateral crown growth of the bordering trees (dashed lines) at sites with different dominant height (LCG$_{17, 21 \& 25}$). For references of the arrows ①, ② and ③ see above.

3.5 Adaptability of GSS to *N. pumilio* natural dynamics

Managing an uneven-aged forest through selection cuts implies a continuous production of wood, so that the remaining stand becomes very important. The regeneration which is established after each harvest and the remaining young trees with timber potential will be the wood source in the coming rotation cycles, so that they constitute the basis for the system´s sustainability (Antequera et al., 1999). That is why post-harvest mortality is a factor of utmost importance.

The harvested stands are affected in their stability, according to the original structure, topography and the type of intervention (Burschel & Huss, 1997, Smith et al., 1997). This weakening effect leads to the fall of trees after the harvest, phenomenon that can seriously affect the quality of the remnant stand. In Tierra del Fuego the windfalls occur even in virgin forests (Rebertus et al., 1997), which poses a logical doubt on the real possibility of implementing this system.

Bava & López Bernal (unpublished data) found that there is no relationship between the manner in which a tree dies (uproot, break or standing death) and the harvest intensity, site quality or stand structure. However, a higher percentage of uprooted trees were observed. The stems that break down correspond to well-anchored individuals, when the wind burden cannot be transmitted by the trunk to the root and soil (Abetz, 1991), or to trees affected by rots, as frequently happens in lenga forests. The uprooting happens when the wind burden is transmitted to the root but cannot be transmitted to the soil (Abetz, 1991). In lenga forests of Tierra del Fuego this can occur in shallow soil stands, when the root system grows superficially (Bava, 1999).

The post-harvest mortality is not significantly related with the percentage of extracted BA. However, when we compare between different stand structures, we note that uneven-aged forests presented minor damage to the even-aged, while the bi-stratified stands presented intermediate damages (Figure 7, ANOVA $p = 0.014$). These differences may have their origin in phenomena observed at two separate scales: in a stand-scale, uneven-aged forests present a more gradual decline of wind speed from the forest canopy up to the understory, allowing a better adaptation mechanics of trees to wind and giving more stability to the whole (Gardiner, 1995). On the other hand, at the individual-level, Wood (Wood, 1995) observed that the tree develops stems only with the resistance needed to support regular wind intensities, growing adaptively. In this way, the increased heterogeneity of uneven-aged stands would provide more opportunities for development of more resistant individuals, which remain after the harvest, and that play a very important role in the stand stability (Burschel & Huss, 1997, Mattheck et al., 1995, Smith et al., 1997). The structural alterations produced by the harvest causes greater exposure of individuals to wind, but in a different way for each one, and would depend on other factors besides the size of the gaps, the h/d value, the felling damages, and homogeneity of the remnant forest.

We have mentioned the importance of the forest stability for sustainability in a selection cuts system, where the productive potential for future interventions is represented by individuals which remain after harvest. In this sense, the results indicate that the post-harvest losses are a limiting factor for the implementation of this system, and which would only be advisable by uneven-aged forests. Moreover, the system success also depends on the conscientiously choosing of the trees to cut, and to carry out the harvest operations carefully. If these conditions are present, the group selection system would be a viable alternative, which would maintain the forest cover, with a cutting cycle of approximately 35 years and extracting a timber volume equivalent to the historical average.

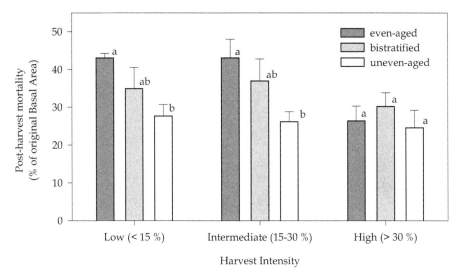

Fig. 7. Post-harvest mortality by harvest intensity and original stand structure. Different symbols indicate significantly different means (Tuckey's posthoc test, $\alpha = 0.05$).

3.6 Case study

In this section we present the main results of three trials located in the province of Tierra del Fuego where group selection cut were applied (Bava & López Bernal, 2006). These were implemented in uneven-aged stands with trees from at least three generations and where it was possible to identify the natural process of gap dynamics.

The tree marking was made in November 2003 and the harvest in February 2004, which consisted of felling and bucking of complete stem. During the tree marking, DBH, height and average sawing bole diameter of all marked trees was recorded. At the same time, it was recorded if the tree was felling to open a new gap, to release existing regeneration, or to optimize the growth of young trees. The felling, skid trails opening and bole extraction were carried out in the same campaign. In all three essays harvest tasks were performed by the same team, using directional felling techniques for tree felling and a skidder for bole extraction.

After the tree marking, a forest inventory was carried out in each of the three trials. Measurements were performed in 300 m² circular plots spread over a 50 m x 50 m grid, representing a sampling intensity of 1.2 %. In each plot, the DBH of all individuals over 10 cm was measured, recording their sawing potential (indicating the length and medium diameter of the logging portion of the bole), and if it had been marked, whether for felling or girdling.

All three trials represented intermediate quality sites, located on gentle slopes and possessing uneven-aged structures. The trial 1 had about 360 tree per ha, a BA of 44 m²/ha and a high proportion of overmature trees (DBH over 60 cm) with a low sawing quality. Essay 2 had 430 trees per ha, a BA of 49 m²/ha and presents a high proportion of trees with a DBH between 40 and 60 cm. Essay 3 had 498 trees per ha, a BA of 52 m²/ha with a high proportion of trees with DBH between 30 and 50 cm (Figure 8).

Timber stock differences between trials derived in great differences on tree marking. The marking intensity, expressed as a percentage of the original AB, was considerably higher in trial 3 than in trial 1 and 2, proportionally to the differences in timber stock. Moreover, differences in the stand structures generated varying amounts of felled and girdled trees (Table 1).

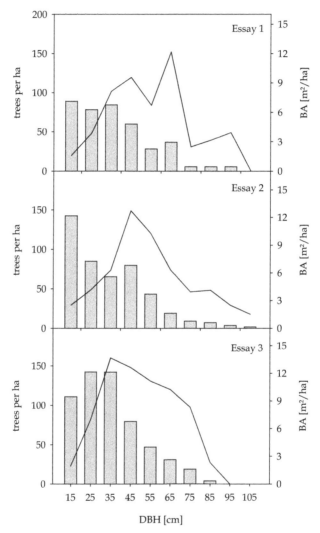

Fig. 8. Diametric frequency distribution for each trial.

The number and size of gaps or patches that were intervened were also different. In the first two trials, which showed similar productions, about 11 gaps per ha were opened by felling or girdling between 2.5 and 3 trees. In trial 3, with a much higher timber production, the number of opened gaps was also bigger, mainly due to a high proportion of patches with young trees (DBH between 20 and 40 cm), while the number of trees per gap increased to 3.4 (Table 2). Moreover, the proportion of gaps or patches with *gap opening*, *gap release* or *patch thinning* interventions differ between essays, pointing out differences in the original stand structures.

Although the three trials were conducted in similar structures, there were significant differences (up to 100%) in the amount of lumber in each. This was reflected in the number of gaps per hectare, but not in their size. The trial with highest harvest intensity (28% of BA) produced twice as sawtimber than the other two, mainly due to felling tending to release young pole trees. This is different from harvests in Chubut province, where the largest volume portion comes from gap opening cuts (Berón et al., 2003).

	Trees (N/ha)			Basal area (m²/ha)		
Essay	felling	girdling	Total	felling	girdling	Total
1	28.0 (87%)	4.0 (13%)	32.0	4.8 (96%)	0.2 (4%)	5.0
2	24.4 (86%)	3.8 (14%)	28.2	4.6 (85%)	0.8 (15%)	5.4
3	58.5 (75%)	19.1 (25%)	77.6	11.8 (81%)	2.8 (19%)	14.6
Mean	37.0 (81%)	9.0 (19%)	45.9	7.1 (86%)	1.3 (14%)	8.3

Table 2. Number and proportion of trees and AB marked in each essay, distinguishing between felling and girdling.

Intervention objective	Essay 1	Essay 2	Essay 3	Mean
Gap opening (N/ha)	2,0	7,6	7,3	6,7
Gap release (N/ha)	1,3	0,4	5,2	2,7
Patch thinning (N/ha)	7,3	3,2	10,2	6,9
Total gaps / patches per ha	10,7	11,2	22,8	16,4
Felled trees per gap	2,6	2,2	2,6	2,5
Girdled trees per gap	0,4	0,3	0,8	0,7
Total	3,0	2,5	3,4	3,1

Table 3. Number of interventions for gap opening, gap release or patch thinning per ha, and number of marked trees per gap in each essay.

According to the remnant structures after harvesting, all three trials are able to recover the volume of extracted timber. However, the best choices to implement a group selection system are stands like in trial 3, i.e. a forest with uneven-aged structure and with a high proportion of trees with DBH between 30 and 50 cm. These structures allow a higher proportion of "gap release" and "patch thinning" interventions, which generates a bigger timber harvest in the first cycle, leaving a high number of young trees in optimal growth conditions. The harvest intensity of this trial is very similar to the historical average for Tierra del Fuego province, at about 27% of BA (Bava & López Bernal, 2004), while is much higher than the historical mean for the province of Chubut, of about 15% (Berón, et al., 2003).

4. Conclusion

The Group Selection System is a valid alternative management system for lenga forests of Argentinean Patagonia. This system emulates one of the most common natural dynamic processes in these forests and provides optimal conditions for regeneration establishment and further development. It is especially recommended for sites with medium to low rainfall levels, where the frequency of large-scale disturbances is low and where the forest presents a natural uneven-aged structure. In Argentina, these situations mainly occur in Chubut province and in the northern part of the lenga distribution in Tierra del Fuego province, where there are already experiences with this type of management.

Moreover, the GSS is compatible with the local production system, dominated by small and medium producers, without financial or technological capacity to afford the costs of intensive harvesting or long-term silvicultural investments. The GSS is adapted to these systems by splitting the turnover age in shorter cutting cycles, giving a more flexible cash flow to these systems, and by allowing that in a single intervention, different silvicultural practices can be carried out. This last point is also an advantage for state control agencies by allowing them to condition the timber extraction to the implementation of other practices that do not generate immediate benefits, such as thinning or regeneration release.

Finally, to ensure the sustainability of forests managed by the GSS, there are at least two aspects that should be especially considered. The first one is that the forester must make his proper interpretation of the natural forest dynamics to decide whether it is feasible or not the implementation this system. The second one implies that to maintain the productive potential for future interventions, logging activities should be conducted with special attention to the remaining forest, using low-impact harvesting technologies.

5. References

Abetz, P. (1991). Sturmschäden aus waldwachstumskundlicher sicht. *AFZ*, Vol. 12, pp. 626-629, ISSN: 0001-1258

Albanesi, E., Gugliotta, O. I., Mercurio, I. & Mercurio, R. (2008). Effects of gap size and within-gap position on seedlings establishment in silver fir stands. *iForest - Biogeosciences and Forestry*, Vol. 1, pp. 55-59, ISSN: 1971-7458

Antequera, S. H., Trhren, M., Bava, J. O., Hampel, H. & Akca, A. (1999). Estudio comparativo de cuatro tratamientos silvícolas en un bosque de lenga de Chubut. *Patagonia Forestal*, Vol. 5, No. 1, pp. 7-10, ISSN: 1514-2280

Arce, J., Peri, P. L. & Martinez Pastur, G. (1998). Estudio de la regeneración avanzada de lenga *Nothofagus pumilio* bajo diferentes alternativas de conducción silvícolas. *Proceedings of Primer Congreso Latinoamericano IUFRO*, Valdivia, Chile, 1998

Barros, V. R., Cordon, V. H., Moyano, C. L., Méndez, R. J., Forquera, J. C. & Pizzio, O. (1983). Cartas de precipitación de la zona oeste de las provincias de Río Negro y Neuquén. Fac. Cs. Agr., Univ. Nac. del Comahue, Cinco Saltos

Bava, J. O. (1999). *Aportes ecológicos y silviculturales a la transformación de bosques vírgenes de lenga en bosques manejados en el sector argentino de Tierra del Fuego*, CIEFAP, ISBN: 1514-2264, Esquel (Argentina)

Bava, J. O., Lencinas, J. D. & Haag, A. (2006). Determinación de la materia prima disponible para proyectos de inversión forestal en la provincia del Chubut. Consejo Federal de Inversiones, pp. 117

Bava, J. O. & López Bernal, P. M. (2004). Análisis de la factibilidad técnica de la aplicación de cortas de selección. Segunda Fase. Consejo Federal de Inversiones, pp. 55

Bava, J. O. & López Bernal, P. M. (2005). Cortas de selección en grupo en bosques de lenga. *IDIA XXI*, Vol. 5, No. 8, pp. 39-42

Bava, J. O. & López Bernal, P. M. (2006). Cortas de selección en grupo en bosques de lenga de Tierra del Fuego. *Quebracho*, Vol. 13, pp. 77-86, ISSN: 0328-0543

Bava, J. O. & Puig, C. J. (1992). Regeneración natural de lenga. Análisis de algunos factores involucrados. *Proceedings of Actas del Seminario de Manejo forestal de la lenga y aspectos ecológicos relacionados*, Esquel, Chubut, Argentina

Bava, J. O. & Rechene, D. C. (2004). Dinámica de la regeneración de lenga (*Nothofagus pumilio* (Poepp. et Endl) *Krasser*) como base para la aplicación de sistemas silvícolas, In: *Ecología y Manejo de los Bosques de Argentina*, Arturi, M. F., Frangi, J. L. & Goya, J. F., pp. 1-22, Editorial de la Universidad Nacional de La Plata, La Plata

Becking, R. W. (1995). Plenterung, an age-old paradigm for sustainability, In: www.ou.edu/cas/botany-micro/ben/, Date of access: 26/07/2011 03:42 p.m., Available from: http://www.ou.edu/cas/botany-micro/ben/ben089.html

Berón, F., Rôo, G. A. & Featherston, S. A. (2003). Los bosques de lenga (*Nothofagus pumilio* (Poepp. et Endl.) Krasser). Su aprovechamiento en la Provincia del Chubut. *Patagonia Forestal*, Vol. 9, No. 2, pp. 14-16, ISSN: 1514-2280

Bridges, L. (2000). *El último confín de la tierra*, Sudamericana, ISBN: 9500718588, Buenos Aires

Brokaw, N. V. L. (1982). The definition of treefall gap and its effect on measures in forest dynamics. *Biotropica*, Vol. 14, No. 2, pp. 158-160, ISSN: 0006-3606

Burgos, J. J. (1985). Clima del extremo sur de Sudamérica, In: *Transecta Botánica de la Patagonia Austral*, Boelcke, O., Moore, D. M. & Roig, F. A., CONICET (Argentina), Royal Society (Gran Bretaña) e Instituto de la Patagonia (Chile)

Burschel, P. & Huss, J. (1997). *Grundriß des Waldbaus. 2 neubearbeitete und erweiterte Auflage*, Berlin-Parey, ISBN: 3-8001-4570-7, Berlin

Callaway, R. M. & Pugnaire, F. I. (2007). Facilitation in plant communities, In: *Functional plant ecology*, Pugnaire, F. I. & Valladares, F., pp. 435-455, CRC Press, ISBN: 978-0-8493-7488-3, Boca Raton

Cavieres, L. A. & Fajardo, A. (2005). Browsing by guanaco (*Lama guanicoe*) on *Nothofagus pumilio* forest gaps in Tierra del Fuego, Chile. *Forest Ecology and Management*, Vol. 204, No. 2-3, pp. 237-248, ISSN: 0378-1127

Cruz M., G. & Schmidt, H. (2007). Silvicultura de los bosques nativos, In: *Biodiversidad: Manejo y conservación de recursos forestales*, Hernández P., J., De la Maza A., C. L. & Cristián, E. M., pp. 279-307, Editorial Universitaria, ISBN: 978-956-11-1969-7, Santiago de Chile

Choler, P., Michalet, R. & Callaway, R. M. (2001). Facilitation and competition on gradients in alpine plant communities. *Ecology*, Vol. 82, No. 12, pp. 3295-3308, ISSN: 0012-9658

Daniel, T. W., Helms, J. A. & Baker, F. S. (1979). *Principles of silviculture* (2nd.), McGraw-Hill, ISBN: 0-07-015297-7, New york

Davis, L. S. & Johnson, K. N. (1987). *Forest Management* (Third), McGraw Hill, ISBN: 0-07-032625-8, New York

Donoso Z., C. (1987). Variación natural en especies de *Nothofagus* en Chile. *Bosque*, Vol. 8, No. 2, pp. 85-97, ISSN: 0304-8799

Donoso Z., C. (1995). *Bosques templados de Chile y Argentina. Variación, estructura y dinámica* (Tercera edición), Editorial Universitaria, ISBN: 956-11-0926-3, Santiago de Chile

Fajardo, A. & McIntire, E. J. B. (2010). Under strong niche overlap conspecifics do not compete but help each other to survive: facilitation at the intraspecific level. *Journal of Ecology*, Vol.99, No. 2, pp. 642-650, ISSN: 1365-2745

Gardiner, B. A. (1995). The interactions of wind and tree movement in forest canopies, In: *Wind and trees*, Coutts, M. P. & Grace, J., pp. 41-59, Cambridge University Press, ISBN: 978-0-521-46037-8, Cambridge

Gea Izquierdo, G., Martinez Pastur, G., Cellini, J. M. & Lencinas, M. V. (2004). Forty years of silvicultural management in southern *Nothofagus pumilio* primary forests. *Forest Ecology and Management*, Vol. 201, No. 2-3, pp. 335-347, ISSN: 0378-1127

Green, P. T. (1996). Canopy Gaps in Rain Forest on Christmas Island, Indian Ocean: Size Distribution and Methods of Measurement. *Journal of Tropical Ecology*, Vol. 12, No. 3, pp. 427-434, ISSN: 0266-4674

Heinemann, K. & Kitzberger, T. (2006). Effects of position, understorey vegetation and coarse woody debris on tree regeneration in two environmentally constrasting forests of north-western Patagonia: a manipulative approach. *Journal of Biogeography*, Vol. 33, No. 8, pp. 1357-1367, ISSN: 03050270

Heinemann, K., Kitzberger, T. & Veblen, T. T. (2000). Influences of gap microheterogeneity on the regeneration of Nothofagus pumilio in a xeric old-growth forest of northwestern Patagonia, Argentina. *Canadian Journal of Forest Research*, Vol. 30, No. 1, pp. 25-31, ISSN: 0045-5067

Jobbágy, E. G., Paruelo, J. M. & León, R. J. C. (1995). Estimación del régimen de precipitación a partir de la distancia a la cordillera en el noroeste de la Patagonia. *Ecología Austral*, Vol. 5, No. 1, pp. 47-53, ISSN: 1667-782X

Kitzberger, T. & Veblen, T. T. (1999). Fire-induced changes in northern Patagonian landscapes. *Landscape Ecology*, Vol. 14, No. 1, pp. 1-15, ISSN: 0921-2973

Lencinas, M., Martínez Pastur, G., Rivero, P. & Busso, C. (2008). Conservation value of timber quality versus associated non-timber quality stands for understory diversity in Nothofagus forests. *Biodiversity and Conservation*, Vol. 17, No. 11, pp. 2579-2597, ISSN: 1572-9710

Lima, R. A. F. d. (2005). Gap size measurement: The proposal of a new field method. *Forest Ecology and Management*, Vol. 214, No. 1-3, pp. 413-419, ISSN: 0378-1127

López Bernal, P. M., Arre, J. S., Schlichter, T. & Bava, J. O. (2010). The effect of incorporating the height of bordering trees on gap size estimations: the case of Argentinean Nothofagus pumilio forest. *New Zealand Journal of Forestry Science*, Vol. 40, pp. 71-81, ISSN: 1179-5395

López Bernal, P. M., Bava, J. O. & Antequera, S. H. (2003). Regeneración en un bosque de lenga (*Nothofagus pumilio* (Poepp. et Endl.) *Krasser*) sometido a un manejo de selección en grupos. *Bosque*, Vol. 24, No. 2, pp. 13-21, ISSN: 0304-8799

Martínez Pastur, G., Lencinas, M. V., Cellini, J. M., Peri, P. L. & Soler Esteban, R. (2009). Timber management with variable retention in *Nothofagus pumilio* forests of

Southern Patagonia. *Forest Ecology and Management*, Vol. 258, No. 4, pp. 436-443, ISSN: 0378-1127

Mattheck, C., Behtge, K. & Albrecht, W. (1995). Failure models of trees and related failure criteria, In: *Wind and trees*, Coutts, M. P. & Grace, J., pp. 195-203, Cambridge University Press, ISBN: 978-0-521-46037-8, Cambridge

Minckler, L. S. & Woerheide, J. D. (1965). Reproduction of Hardwoods 10 Years After Cuttting as Affected by Site and Opening Size. *Journal of Forestry*, Vol. 63, No. 7, pp. 103-107, ISSN: 0022-1201

Musters Chaworth, G. (1871). *Vida entre los Patagones*, Solar, Buenos Aires

Mutarelli, E. & Orfila, E. (1971). Observaciones sobre la regeneración de lenga, *Nothofagus pumilio* (Poepp. et Endl.) *Oerst.*, en parcelas experimentales del Lago Mascardi, Argentina. *Revista Forestal Argentina*, Vol. 15, No. 4, pp. 109-115

OMNR. (2004). Ontario Tree Marking Guide, Version 1.1. Ont. Min. Nat. Resour. Queen's Printer for Ontario, pp. 252

Otero Durán, L. (2006). *La huella del fuego. Historia de los bosques nativos. Poblamiento y cambios en el paisaje del sur de Chile*, CONAF - Pehuén Editores, ISBN: 9561604094, Santiago, Chile

Perera, A. H., Buse, L. J., Weber, M. G. & Crow, T. R. (2004). Emulating natural disturbance in forest management: a synthesis, In: *Emulating natural disturbance in forest management: an overview*, Perera, A. H., Buse, L. J. & Weber, M. G., pp. 3-7, Columbia University Press, ISBN: 9780231129176, New York

Popma, J., Bongers, F., Martinez-Ramos, M. & Veneklaas, E. (1988). Pioneer Species Distribution in Treefall Gaps in Neotropical Rain Forest; A Gap Definition and Its Consequences. *Journal of Tropical Ecology*, Vol. 4, No. 1, pp. 77-88, ISSN: 0266-4674

Puettman, K. J., Coates, K. D. & Messier, C. (2009). *A critique of silviculture: managing for complexity* (1º), Island Press, ISBN: 978-1-59726-146-3, Washington

Rebertus, A. J., Kitzberger, T., Veblen, T. T. & Roovers, L. M. (1997). Blowdown history and landscape patterns in the Andes of Tierra del Fuego, Argentina. *Ecology*, Vol. 78, No. 3, pp. 678-692, ISSN: 0012-9658

Rebertus, A. J. & Veblen, T. T. (1993). Structure and tree-fall gap dynamics of old-growth *Nothofagus* forests in Tierra del Fuego, Argentina. *Journal of Vegetation Science*, Vol. 4, No. 5, pp. 641-654, ISSN: 1100-9233

Rosenfeld, J. M., Navarro Cerrillo, R. M. & Guzman Alvarez, J. R. (2006). Regeneration of *Nothofagus pumilio* [Poepp. et Endl.] Krasser forests after five years of seed tree cutting. *Journal of Environmental Management*, Vol. 78, No. 1, pp. 44-51, ISSN: 1365-2745

Runkle, J. R. (1981). Gap regeneration in some ald-growth forests of the eastern United States. *Ecology*, Vol. 62, No. 4, pp. 1041-1051, ISSN: 0012-9658

Runkle, J. R. (1985). Disturbance regimes in temperate forest, In: *The ecology of natural disturbance and patch dynamics*, Pickett, S. T. A. & White, P. S., pp. 17-33, Academic Press Inc., ISBN: 978-0125545211, Orlando

Rusch, V. (1992). Principales limitantes para la regeneración de la lenga en la zona N.E.de su área de distribución. *Proceedings of Actas del Seminario de Manejo forestal de la lenga y aspectos ecológicos relacionados*, Esquel, Chubut, Argentina,

Schlatter, J. E. (1994). Requerimientos de sitio para la lenga, *Nothofagus pumilio* (Poepp. et Endl.) Krasser. *Bosque*, Vol. 15, No. 2, pp. 3-10, ISSN: 0304-8799

Smith, D. M., Larson, B. C., Kelty, M. J. & Ashton, P. M. S. (1997). *The practice of silviculture. Applied forest ecology* (9° ed.), John Willey & Sons, ISBN: 0-471-10941-X, New York

Tortorelli, L. A. (2009). *Maderas y bosques argentinos* (2° ed), Editorial ACME, ISBN: 978-987-9260-69-2, Buenos Aires, Argentina

Veblen, T. T., Ashton, D. H., Schlegel, F. M. & Veblen, A. T. (1977). Plant Succession in a Timberline Depressed by Vulcanism in South-Central Chile. *Journal of Biogeography*, Vol. 4, No. 3, pp. 275-294, ISSN: 03050270

Veblen, T. T. & Donoso, C. (1987). Alteración natural y dinámica regenerativa de las especies chilenas de *Nothofagus* de la región de los lagos. *Bosque*, Vol. 8, No. 2, pp. 133-142, ISSN: 0304-8799

Veblen, T. T. & Donoso Z., C. (1987). Alteración natural y dinámica regenerativa de las especies chilenas de *Nothofagus* de la Región de Los Lagos. *Bosque*, Vol. 8, No. 2, pp. 133-142, ISSN: 0304-8799

Veblen, T. T., Donoso Z., C., Kitzberger, T. & Rebertus, A. J. (1996). Ecology of southern chilean and argentinian *Nothofagus* forests, In: *The ecology and biogeography of Nothofagus forests*, Veblen, T. T., Hill, R. S. & Read, J., pp. 293-353, Yale University Press, ISBN: 0-300-06423-3, London

Veblen, T. T., Donoso Z., C., Schlegel, F. M. & Escobar R., B. (1981). Forest dynamics in South-central Chile. *Journal of Biogeography*, Vol. 8, No. 3, pp. 211-247, ISSN: 03050270

Veblen, T. T., Schlegel, F. M. & Escobar R., B. (1980). Structure and dynamics of old-growth *Nothofagus* forests in the Valdivian Andes, Chile. *Journal of Ecology*, Vol. 68, No. 1, pp. 1-31, ISSN: 1365-2745

Willis, B. (1914). The Physical Basis of the Argentine Nation. *The Journal of Race Development*, Vol. 4, No. 4, pp. 443-460, ISSN: 10683380

Wood, C. J. (1995). Understanding wind forces on trees, In: *Wind and trees*, Coutts, M. P. & Grace, J., pp. 133-164, Cambridge University Press, ISBN: 978-0-521-46037-8, Cambridge

York, R. A., Heald, R. C., Battles, J. J. & York, J. D. (2004). Group selection management in conifer forests: relationships between opening size and tree growth. *Canadian Journal of Forest Research*, Vol. 34, No. 3, pp. 630-641, ISSN: 0045-5067

Zhu, J., Hu, L., Yan, Q., Sun, Y. & Zhang, J. (2009). A new calculation method to estimate forest gap size. *Frontiers of Forestry in China*, Vol. 4, No. 3, pp. 276-282, ISSN: 1673-3630

Case Study of the Effects of the Japanese Verified Emissions Reduction (J-VER) System on Joint Forest Production of Timber and Carbon Sequestration

Tohru Nakajima
Laboratory of Global Forest Environmental Studies
Graduate School of Agricultural and Life Sciences
The University of Tokyo, Yayoi, Bunkyo-ku
Laboratory of Forest Management
Graduate School of Agricultural and Life Sciences
The University of Tokyo, Yayoi, Bunkyo-ku, Tokyo
Japan

1. Introduction

In the context of climate change (including global warming), the net reduction in carbon emissions as a result of forest carbon sinks and sustainable forest management are two critical issues. Recently, the benefits of carbon sequestration by forests have been highlighted and carbon sequestration has been measured throughout the world: in the United States (Sakata 2005; Calish et al 1978; Foley et al 2009; Ehman et al 2002; Im et al 2007), Europe (Backèus et al 2005; Liski et al 2001; Matala et al 2009;Pohjola and Valsta 2007;Sivrikaya et al 2007; Kaipainen et al 2004; Seidl et al 2007), Canada (Hennigar et al 2008; Thompson et al 2009), Oceania (Campbell and Jennings 2004) and Asia (Ravendranath 1995; Han and Youn 2009). Forests not only have economic value through the production of commercial timber, but they also have other values to society including acting as carbon sinks, supporting biodiversity, and providing water protection (Pukkala 2002). Forest management subsidies are required from national budgets (funded by the tax payer) to increase the public benefit of forests by restricting the area that is clear cut and preventing other damaging silvicultural activities from being practiced. On the other hand, in the absence of artificial thinning, intensive self-thinning can occur (Nakajima et al., 2011d), resulting in significant CO_2 emissions. Therefore, both thinning and harvesting are necessary not only for commercial timber production, but also in order to reduce CO_2 emissions and gain carbon credits. In addition, tree growth gradually decreases with age (Nakajima et al., 2010; 2011d; Pienaar and Turnbull 1978), so older stands will eventually cease to increase their carbon stock. It, therefore, makes economic sense to undertake clear felling before such stagnation occurs in older stands and carbon credits are no longer available.

Because Japanese forest management profits have been in decline as a result of lower timber prices (Forestry Agency 2007), almost all Japanese forest owners depend on government subsidies to maintain their forests (Komaki 2006; Nakajima et al 2007b). Previous studies have shown that the area of silvicultural practice including planting, weeding, pruning, pre-commercial thinning and thinning, is strongly correlated with the amount of national subsidy that is provided (Hiroshima and Nakajima 2006). Therefore, the planted forests of Japan that are funded by national subsidies should be in a condition suitable for the public to benefit from them. Generally, it is not possible to rely on natural regeneration in planted forests in Japan. The silvicultural practices used to ensure regeneration in Japan have been described in previous studies (Nakajima et al., 2011b; Sakura 1999; Ohtsuka, 1993) and are outlined in table 1.

Silvicultural practices	Stand age (year)
Land preparation and planting	0
Weeding	$1-10^*$
Pruning	15
Precommecial thinning	20, 25

* Weeding is undertaken every year in stands aged 1 to 10 years

Table 1. Silvicultural practices undertaken at the study site

In addition, Japanese citizens think that acting as carbon sinks will be one of the most important functions of forests in the future (Forestry Agency 2007). Based on public opinion, it would be an valuable for forest managers to include the carbon benefits in their forestry profit predictions.

In order to include carbon benefits in forest management, a number of previous studies have proposed what is known as the 'social rule' (Im et al 2007; Foley et al 2009; Hennigar et al 2008). Because the rotation period is important for forest management decision making and strongly affected by regional forest resources, some studies have focused on estimating how the optimum rotation period is affected by different carbon offset systems. Carbon offsetting may be advantageous for forest management based on optimizing the rotation period (Raymer et al 2009), but it can be disadvantageous because of the effects of natural disturbance, which can release carbon (Galik and Jackson 2009). However, few studies have investigated the effects of existing carbon offsetting programs (including forest carbon sinks) in the context of global warming policy frameworks.

Under the global policy framework (resulting from the Kyoto Protocol) the size of the carbon sink in a forest is calculated for forests that have experienced afforestation, reforestation and deforestation (ARD forests) since 1990, as described by Article 3.3; and in terms of forests where silvicultural practices have been conducted since 1990 (FM forests) under Article 3.4.

The Kyoto Protocol requires signatories to reduce their CO_2 emissions and other greenhouse gases by their quantified reduction commitments below 1990 levels during the first commitment period (FCP), 2008–2012. Now that the end of the FCP is fast approaching, each country is preparing to report on emissions and the removal of carbon by forests in accordance with the Good Practice Guidance for Land Use, Land-Use Change, and Forestry (GPG-LULUCF) (Amano 2008a; Houghton et al 1997; IPCC 2000; IPCC 2007). In the protocol, Japan is committed to reducing CO_2 equivalent emissions to 6% below its 1990 level (Amano 2008b; Amano and Tsukada 2006). At the same time, the protocol allows net changes in greenhouse gas emissions to be included. For example, removal by sinks resulting from human-induced land-use changes and forestry (LUCF) activities can be added to or subtracted from their reduction commitments as appropriate.

Under the Kyoto mechanism, carbon emission trading can be undertaken. Carbon dioxide (CO_2) credits have already been traded in some markets, such as the carbon market in the United Kingdom since April 2002. The carbon price is expected to affect forestry profits and has the potential to cause considerable changes to harvesting ages. Predicting how changes in the cutting age affect carbon prices encourages the consideration of forestry measures in these terms. In order to quantify carbon storage, previous studies have proposed various methods for estimating carbon credits, including the stock changing, average storing, and ton-year methods (Richards and Stokes 1994; Schroeder 1992; Moura-Costa and Wilson 2000).

To accelerate efforts to combat global warming in accordance with the Kyoto Protocol, based on the stock changing method, Japan's Ministry of the Environment has established a forest carbon credit system. The system, which is based on the Japan Verified Emissions Reduction (J-VER) system, was launched in November 2008 and will help in calculations of forest CO_2 absorption. This is the first system of its kind. The absorption will be calculated in credits, which can then be sold to CO_2-emitting companies already registered in the J-VER system. The Ministry hopes that the credits will be traded on the carbon market in the future and funds reinvested in the expansion of the current area where silviculture is practiced; this can then be counted as CO_2 absorption under the Kyoto protocol.

Carbon accounting is based on accounting systems developed as part of the FCP under the Kyoto Protocol. Three project types are particularly important in the J-VER system: thinning promotion and management; sustainable forest management; and plantation management. Areas thinned after 2007 will be the target of the efforts under the Japanese system. Sustainable forest management activities will focus on areas that were harvested and replanted after 1990. Plantation projects will focus on replanting, and all forests eligible for credits under the credit system need to have a forest management system compliant with current Forest Law.

No previous study has clarified the effect of this new carbon offset accounting system on the actual forest area formally identified in the J-VER system. For medium- to long-term forest management strategies, it is important to clarify the effect of the J-VER system on forestry strategies. Therefore, this study aimed to investigate the effects of the carbon offsetting system on the carbon stock and timber production relative to the carbon price. Because harvesting activities need to be included in long-term forest management, we examined the sustainable forest management project under the J-VER system.

2. Materials and methods

2.1 Study area

This research was conducted in the University of Tokyo Forest, located in the cities of Kamogawa and Kimitsu, Chiba Prefecture, Japan (Fig. 1).

This forest lies 50 to 370 m above sea level and is characterized by undulating terrain with steep slopes and primarily brown forest soils. It is located in a warm temperate zone, with an average annual temperature of 14°C and an average annual precipitation of 2182 mm. The total forest area is 2216 ha; 824 ha (37%) contain sugi (*Cryptomeria japonica*) and hinoki (*Chamaecyparis obtusa*) stands, 949 ha (43%) are natural hardwood forest, and 387 ha (17%) are natural conifer forest. The remaining 57 ha (3%) are occupied by a demonstration forest. Many permanent research plots have been established in sugi stands within the forest since

1916, and tree height, height to crown base, and diameter at breast height (DBH) have been recorded approximately every 5 years since that time. A national subsidy system for the thinning of all planted tree species is commonly applied, but mainly to forest plantations less than 35 years old. The grant rates of the subsidy systems cover approximately 70 % of the cost of thinning. Inventory data relating to the private forests, such as stand age, area, tree species, slope, address of forest owners and site index, were available and were also linked to each stand included in the geographic information system (GIS). Using the inventory data, age distribution at this study site was derived and is shown in Figure 2. The site index map in this study area was also established using the airborne LiDAR measurements (Hirata et al., 2009; Hiroshima and Nakajima 2009). Only sugi (*Cryptomeria japonica*), the best-known planted tree species in Japan, was considered.

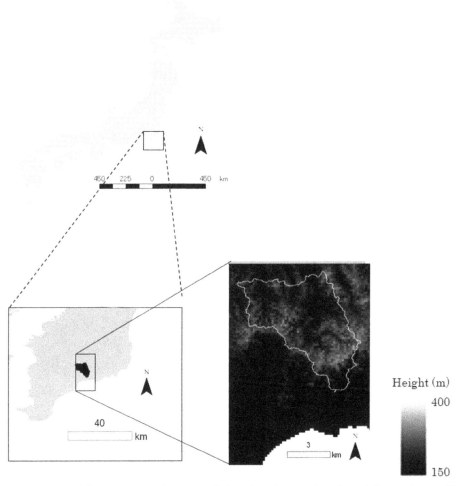

Fig. 1. Location of the University Forest in Chiba, showing an elevation of the study site. The blue line shows the forest boundary line of the University Forest in Chiba.

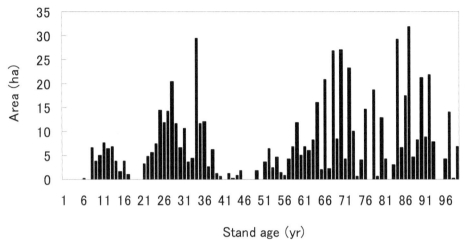

Fig. 2. The age distribution of forested areas in the study site

Approximately 58 % of planted forests in Japan are privately owned (Forestry Agency, 2007), and the forest policy subsidy system is known to have a great influence on the management practiced within them (Hiroshima and Nakajima, 2006). Furthermore, due to the socio-economic situation in Japan, there has been little financial incentive to practice sound forest management and profits have been very low as a result of decreasing timber prices. This has resulted in increased areas of unmanaged and unthinned forests, many of which have been left untended for more than 10 years (Nakajima et al., 2007).

Hence, there is an urgent need to improve the profitability of Japanese forestry. Due to the general lack of thinning, self-thinning has been increasing, accompanied by reductions in the carbon stock and adverse effects on forest ecosystem functioning. These developments are in direct conflict with a need to increase thinned areas of forest, relative to 1990 levels (Japanese Forestry Agency, 2007), under Kyoto Protocol commitments (Houghton et al., 1997; UNFCC, 1998; Robert et al., 2000; UNFCC, 2002; IPCC, 2003; Jansen and Di, 2003). Thus, there is an urgent need to expand the areas that are subject to planned thinning, and to reduce the cost of such operations by increasing their scale through forest owner cooperation. Therefore, silvicultural practices are now supported by a subsidy system (Nakajima et al., 2007), under which forest owners are required to report the conditions of their stands and the silvicultural treatments they have applied. The central government and local Prefectural government subsidize the thinning of planted forests containing trees younger than 35 years of any species, meeting approximately 70 % of the thinning cost. The subsidies for thinning are available in forests that have been subsidized in the preceding five years.

This area was one of the forest projects formally identified in Japan's Verified Emission Reduction system (J-VER), which is a Japanese carbon offset system. It is important, therefore, to establish a sustainable forest management system that takes into consideration timber production and the amount of carbon stock held in the area.

2.2 Analysis tool

The data source and analysis tool used in this study for estimating carbon absorbed by the forest were developed in accordance with the J-VER guidelines (Environmental Ministry

2009), which are based on the carbon accounting system developed for the Kyoto Protocol. J-VER guidelines suggest the use of the Local Yield Table Construction System (Nakajima et al. 2009a; Nakajima et al. 2010), which is a timber growth and carbon stock simulator. This growth model is applicable to the main tree species, including sugi (*Cryptomeria japonica*), hinoki (*Chamaecyparis obtusa*), karamatsu (*Larix leptolepis*) and todomatsu (*Abies sachalinensis*), which are planted throughout Japan. By combining LYCS with a wood conversion algorithm and a harvesting cost model (Nakajima et al. 2009a; 2009c), we can predict not only carbon stock but also harvested timber volume and forestry income. The stand age and tree species included in the forest inventory data can be used as input data for the LYCS. The harvest and silvicultural practice records of the study site, including details of incomes, costs, and labor, were used to estimate forestry profits for harvesting and silviculture. The unit price of subsidies depends on the standard silviculture system and historical records of the amount of labor required to carry out various silvicultural practices including silviculture treatments (planting, weeding, pruning, pre-commercial thinning) and harvesting (thinning, clear-cutting) were also available from the University forest in Chiba.

2.3 Data analysis

In the present study, we investigate through simulation modeling the effects of the J-VER system on timber production, carbon stock holdings. Two carbon price scenarios were assumed: Scenario 1 was no J-VER system applied to stands; Scenario 2 was the J-VER system fixing the carbon price to 1000 yen/ton-CO_2 considering previous research (Nakajima et al. 2011c), applied to stands. The international pledge made under the Kyoto Protocol commitments (Houghton et al., 1997; UNFCC, 1998; UNFCC, 2002), requires a 6 % reduction of CO_2 emissions from the 1990 level, of which 3.8 % may be attributed to carbon absorption by means of 'forest management' (Hiroshima 2004; Forestry Agency 2007). Increasing the area of 'forest management' as described under article 3.4 in the Kyoto Protocol, requires pre-commercial or commercial thinning (Nakajima et al., 2007a). Therefore, to fulfill Japan's international pledge under the Kyoto Protocol in a global context (Hiroshima and Nakajima et al., 2006), it has been proposed that a new J-VER system (i.e. Scenario 2) can be applied. This will promote thinning and restrict large-scale clear cutting by supporting long-rotation silviculture (Forest Agency 2007).

Based on the assumptions of the two scenarios, the harvesting area, amount of harvested timber, subsidy, forestry profits, carbon stock and quantity of labor were calculated by using an existing stand growth model (Nakajima et al. 2010), a wood conversion algorithm (Nakajima et al. 2009c) and a forestry cost model (Nakajima et al. 2009a). With data describing the stand condition (stand age, site index and tree species), the thinning plan (thinning ratios, number of thinnings and the thinning age) and the timber price as model inputs, the future stand volume, timber volume and forestry profits can be generated as model output (Nakajima et al., 2009a, 2009c, 2010).

The accuracy of the basic model for predicting future stands has been exhaustively checked by comparing estimated tree growth with observed tree growth data in permanent plots (Ohmura et al. 2004) gathered over more than 30 years (Shiraishi 1986; Nakajima et al. 2010).

By inputting the stand condition into these models, the future forestry profits could be estimated as a function of the harvesting plan strategies and the carbon price. However, because it is not easy to predict inflation and timber price fluctuations precisely, we assume in the model that the socio-economic situation driving these variables is constant. We therefore assume that timber price remains constant throughout the prediction period and is as described by a previous study (Nakajima et al. 2009a). We believe this assumption is justified since a survey by the forest association, and government reports (Forestry Agency 2007) indicate that the current annual average timber price has been stable over recent years. The final age at cutting was chosen to maximize the present net value of forestry profits, estimated from those valid at the most recent final cutting. Although the thinning plan is included in the input data as mentioned above, it can be changed according to a particular stand density control strategy. The optimum thinning plan was decided upon by selecting the one which maximized the net present value. We varied the thinning ratios by 5 % increments from 20 % to 40 % in line with the existing standard silviculture systems (Forestry Agency 2007). We also varied the number of thinnings between zero and three, and the thinning age by increments of 5 years between the initial stand age and the final age at cutting. By inputting these various thinning plans into the LYCS, we simulated forestry profits under all harvesting strategies. We then selected the cutting plan that maximized the present net value of forestry profits.

The forestry profits could then be estimated from the forestry income and the carbon credit. Sakata (2005) examined the effects of the carbon market on forestry profits in the USA. At the study site selected by Sakata (2005), both saw logs and pulp wood were considered to contribute to any profits. On the other hand, production of pulp wood at the current study site is not commercially viable because the cost of harvesting is so high. Therefore, the study described herein examined the effects of the carbon market on forestry profits when producing saw logs alone and not pulp wood.

The carbon stocks were also estimated by substituting stand volumes derived from LYCS into the following formula (Environmental Ministry 2009):

$$C = V \cdot D \cdot BEF \cdot (1 + R) \cdot CF \tag{1}$$

where C is the carbon stock (t-C), V is the stand volume (m^3), D is the wood density (t-dm/m^3), BEF is the biomass expansion factor, R is the ratio of below ground biomass to above ground biomass and CF is the carbon content (t-C/t-dm).

The biomass expansion factor for trees younger than 20 years was 1.57; the biomass expansion factor for trees older than 20 years was 1.23; the ratio of below ground biomass to above ground biomass was 0.25; the wood density (tonnes/m^3) was 0.314; and the carbon content (t-C/t-dm) was 0.5 (Environmental Ministry 2009; Fukuda et al. 2003). By multiplying 3.67(44/12=molecule of CO_2/molecule of C) by the amount of the carbon stock present, the amount of CO_2 can be calculated. The carbon credit can be calculated by multiplying the CO_2 increase per year by the carbon price (yen/ton).

Many previous studies (van Kooten et al. 1995; Nakajima et al., 2011c) used increases in timber volume as a base from which to calculate carbon credits. The gain in carbon credits has been calculated on the basis of timber growth, and the release of carbon credits occurred when timber was harvested. In the J-VER system, however, the accounting is based on the total volume of the tree stock (we refer to this method as J-VER accounting). Therefore, when estimating carbon credits under the J-VER system, there is no need to undertake lifecycle assessments. We conducted a sensitivity analysis, in order to clarify the effects on the net present value (NPV) of changes in various parameters, including the initial stand age (0, 20 or 40 years), the site index and the carbon price (CP) and discount rate within the J-VER system.

The traditional final cutting age in order to maintain the maximum mean growth rate in Japanese planted forests is approximately 50 years (The Tokyo University Forests, 2006). Using this age as a reference, we set the initial stand ages in our models to be 0, 20 and 40 years. The discount rate was then estimated relative to a value considered to be reasonable to society; in this case 3.0 % was considered reasonable as this represents the average long-term yield of Japanese government bonds (Tokyo Stock Exchange 2007). Using the discount rate (3.0 %) as a reference, we set the discount rate to 0, 20 and 40 % in our models. Using the yield table presented by Nakajima et al. (2010) as a reference, we set the site indexes 1, 2 and 3 to represent good, intermediate and poor site quality, respectively. We examined various combinations of the different parameters to estimate the NPV of timber production, carbon credits and total NPV. In addition, wind hazard probability is an important parameter; wind it the main natural disturbance in Japanese mountain forests and it increases with increasing stand age and height (Nakajima et al., 2009b; Tsuyuki et al., 2011). The probability of wind disturbance, thus, also increases with time. However, tremendous wind disturbance records were not observed in the study site, so we did not include this parameter when calculating NPV for the forest area studied.

Based on the methodologies for calculating NPV mentioned above, the predictions at the forest level could then be estimated by summarizing the predicted values at the stand level. Because the period of validation over which these previous studies were conducted was longer than the prediction period of 25 years adopted in the present study, estimates of future timber production and forestry profits (Nakajima et al. 2009a; 2009c) could be calculated based on predictions of future tree growth at the level of stands. If the predicted values derived from existing models at the stand level are accurate, it follows that the predicted value at the forest level, which is the sum of values at the stand level, would be also accurate. For descriptive purposes, the prediction period was set to 25 years, which is the period specified for natural resource predictions by the Japanese Ministry of Education, Culture, Sports, Science and Technology (Science Council 2008).

By inputting the stand condition derived from forest inventory data into our models, future forestry profits could be estimated as a function of the harvesting plan strategies and the carbon price. As mentioned above, the discount rate was then estimated relative to a value considered to be reasonable to society; in this case 3.0 % was considered reasonable as this represents the average long-term yield of Japanese government bonds (Tokyo Stock Exchange 2007). The total harvesting area and the quantity of harvested timber were calculated by summarizing their respective values based on the harvesting plans calculated for each of the two scenarios under the carbon price of 0 and 1000 yen/CO_2-ton. The subsidies were estimated by summarizing the silviculture and thinning subsidies derived from government subsidy unit prices. In this study, the term "thinning subsidies" refers to subsidies associated with commercial thinning. In other words, the harvesting is not conducted as part of the silvicultural practices that include pre-commercial thinning. The total forestry profits could then be estimated from the forestry income and the subsidy. The carbon stocks were also estimated by substituting stand volumes derived from LYCS into the following formula (1):

In addition, labor requirements were calculated by multiplying the amount of labor required per hectare for each silvicultural practice, by the area over which that silviculture would be practiced, based on the estimated harvesting plans and the age distribution of trees in the study site.

3. Results and discussion

Results of the sensitivity analysis, based on the initial stand age (0, 20 or 40 years), and taking into account carbon price (CP), discount rate and site index within the J-VER system, are presented in Figure 3.

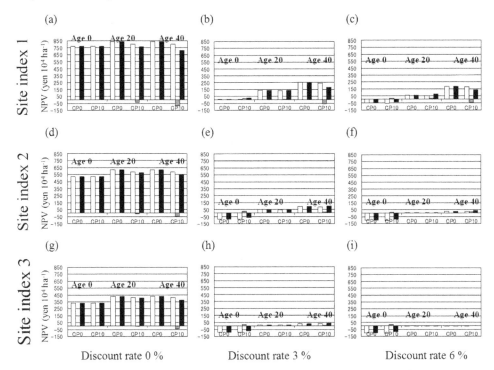

Fig. 3. Sensitivity analysis separated on the basis of initial stand age (0, 20 and 40 years), taking into account the site index, carbon price (CP) and discount rate within the J-VER system.

The white, grey and black bars show, respectively, the NPV of timber production, the carbon credit and the total NPV.

The profits change depending on the site quality, initial stand age, the carbon credit and the discount rate. As shown in figure 3, the higher the initial stand age, the lower the effect of carbon credit on the total profit. In addition, the higher the discount rate, the lower the profit. It is particularly noteworthy that the profit when the initial stand age is 20 years under the J-VER system shown in figure 3b is almost 0. This means that the carbon credit of 1000 yen for a stand with a site index of 1 and an initial stand age of 20 years could be sufficient to compensate landowners and make carbon storage economically attractive.

Several previous studies that have examined the effect of carbon price and taxes on forest management have accounted for carbon stock and release on the basis of timber volume (we call this method 'timber-based accounting' Nakajima et al., 2011c; van Kooten et al. 1995). The J-VER system had a greater impact on forestry profits than the timber-based accounting system (Nakajima et al., 2011c). Generally, the economic effect on the NPV calculated by the

J-VER accounting system was more sensitive than that calculated in previous studies using timber-based accounting (Nakajima et al., 2011c). We consider that the main reason for this result is that the estimated number of carbon credits under the J-VER accounting system is greater than estimates using the timber-based method (Nakajima et al., 2011c; van Kooten et al. 1995). This difference affected the profits derived from different forests depending on the age distribution under the carbon offsetting system. Figure 3 shows the positive or negative effect of stand age and carbon price in the targeted forest area on forestry profits. A strong positive effect was found for younger stands and a negative effect was found for older stands under the J-VER system. For example, in figure 3, the total effect of a carbon price of 1000 yen on the forestry profits for a stand with an initial age of 0 years (e.g. Fig. 3b, e, h) was positive, but with an initial stand age of 40 years (e.g. Fig. 3a, b, c, d, g) the owner would make a loss. Therefore it might be more important to consider stand age distribution, allocation of the harvesting area and carbon price fluctuation when planning forest management under the J-VER accounting system. Under the J-VER system, the total carbon storage included leaves, branches and roots, which were all counted as carbon sinks. Therefore, the lost of carbon credit by emission derived from clear cutting was greater than that calculated using the timber-based accounting system. In general, the age of existing Japanese planted forest stands is increasing (Forestry Agency 2007). Therefore, we suggest that the J-VER system may have a negative effect on forest profitability throughout Japan.

In particular, such negative impacts are likely to be greater in stands with high site quality. Therefore, harvesting, and particularly clear cutting, of stands on high quality sites will decline (Fig.3a-c). At the forest level, for the whole area examined in this study, the harvested area was calculated by summing the stand level harvesting area. Thus, the harvested area at the forest level also decreased under the J-VER system (Fig. 4).

In particular, such negative impacts are likely to be greater in stands with high site quality. Therefore, harvesting, and particularly clear cutting, of stands on high quality sites will decline (Fig.3a-c). At the forest level, for the whole area examined in this study, the harvested area was calculated by summing the stand level harvesting area. Thus, the harvested area at the forest level also decreased under the J-VER system (Fig. 4).

Figure 4 shows: (a) the age distribution of the final cutting area under different scenarios and (b) age class graphs for the scenarios at the end of the 25-year simulations. The former shows that the average stand age, under Scenarios 1 and 2, at the time of clear-cutting was 65 years and 80 years, respectively. The age classes at clear-cutting ranged from 8–15 years under Scenario 1, and from 8–20 years under Scenario 2. Because the target tree species was the most commonly planted species for timber production in Japan (Forestry Agency, 2007), this tendency for a reduction in the harvested area at the forest level studied could be applied to the regional level.

The increase in the potential harvesting area is derived from the increasing area of mature forest as the age distribution of stands in the study site changes over time (Fig. 5).

Under Scenario 1, profits from stands in an age class greater than 4 (36 years old) could be derived from harvest income alone, while under Scenario 2 profits could be derived from harvesting income and carbon sequestration. A comparison of the two scenarios clearly reveals a larger clear-cutting area under Scenario 1 than under Scenario 2 in the initial stage under the prediction period, the difference ranging between 1 ha and 27 ha. In 2021, the magnitude of the difference in clear-cutting areas decreased by up to 5.3 % of its maximum value. In contrast, the thinning area under Scenario 2 is clearly larger than under Scenario 1, with the difference ranging between 10 and 17 ha. These results show that the harvesting practices under the scenarios 1 and 2 were mainly clear cutting and thinning, respectively.

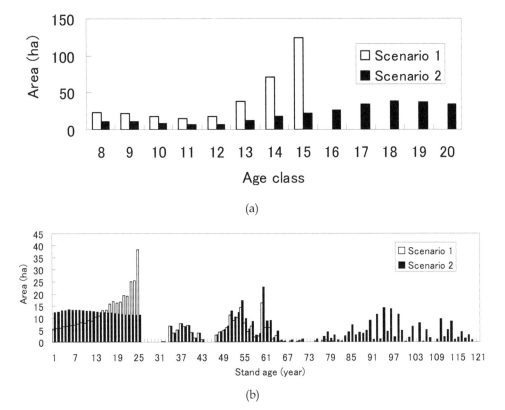

Fig. 4. (a) The age distribution of final cutting area under different scenarios and (b) age class graphs of scenarios at the end of the 25-year simulations. White and black blocks show the final cutting area under Scenarios 1 and 2, respectively.

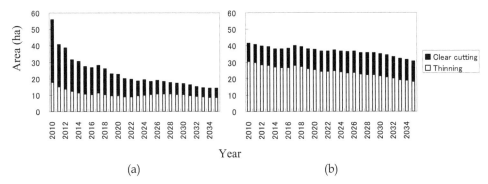

Fig. 5. The clear-cutting and thinning harvesting areas under (a) Scenario 1 and (b) Scenario 2.

White and black blocks show the thinning and clear-cutting harvesting areas, respectively.

3.1 Timber production

Figure 6 shows the differences in volumes of harvested timber under the two scenarios. Under Scenario 1, the harvest of clear-cut timber at the initial stage of the prediction period was larger than that of thinned timber, with a percentage clear-cut to thinned timber ranging from 87 % and 13 % in 2010 to 64% and 36 % in 2033.

After 2011, the volume of harvested timber decreased by up to 15.5 % of its maximum value due to a decrease of harvesting area (Fig. 5a) for clear-cutting.

Under Scenario 2 the clear–cut timber harvest was little larger than that of thinned timber with the percentages of the clear-cut to thinned timber ranging between 47% and 53 % in 2010 to 29% and 71 % in 2035.

The harvested timber volume decreased by up to 91.2 % of its maximum value between 2010 and 2035 due to a reduction in the harvested area (Fig. 5b). Although the total volume of harvested timber under Scenario 1 was larger than that under Scenario 2 up to 2014, in 2015 the pattern was reversed.

A comparison of the two scenarios clearly shows that the harvested volume of clear-cut timber in the initial stage of the prediction period was larger under Scenario 1 than Scenario 2, with differences ranging between 7.7 and 0.3×10^3 m^3. After 2010, the difference between volumes of clear-cut timber decreased by up to 2.5 % of its maximum value. In contrast, the volume of thinned timber harvested under Scenario 2 was clearly larger than under Scenario 1, with differences ranging between 1.2 and 1.6×10^3 m^3. These results show that production was predominantly of clear-cut timber especially under Scenario 1. Comparing Figs 5 and 6 shows that the ratio of clear-cut timber to total harvested timber is higher than the ratios of their respective harvested areas indicating that the volume of harvested timber per unit of harvested area was larger for clear-cut timber than thinned timber. Under the J-VER system (scenario 2), the amount of timber derived from clear cutting, which generally yields timber of larger dimensions than that derived from thinning, would be less than that under the non J-VER system (see figure 6). In particular, in the short-term, the total timber yield would be reduced under the J-VER system.

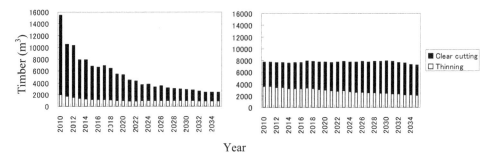

Fig. 6. The clear-cutting and thinning harvested timber volume under (a) Scenario 1, and (b) Scenario 2.

White and black blocks show the thinning and clear-cutting harvested timber volume, respectively.

3.2 Carbon stock

Figure 7 shows the response of the carbon stock to the different scenarios. Under Scenario 1 the maximum and minimum carbon stocks were 49948 tonnes in 2010 and 27639 in 2023. The carbon stock decreased by up to 55.3 % of its maximum due to the reduction in area harvested by clear-cutting (Fig. 5a).

Under Scenario 2 the maximum and minimum carbon stocks were 78037 tonnes in 2010 and 48342 in 2035. Between 2010 and 2035 carbon stock increased by up to 61.9 % of its minimum due to forest growth (Fig. 7b). The total carbon stock was smaller under Scenario 1 than under Scenario 2 throughout the prediction period.

Generally, the carbon stock under Scenario 2 was relatively more stable than that under Scenario 1. A comparison of the two scenarios clearly shows the carbon stock under Scenario 1 to be smaller than under Scenario 2 with differences ranging between 0 and 658.3 Kt suggesting that differences in carbon stock between the two scenarios were mainly due to clear-cutting. According to the carbon accounting system under the Kyoto Protocol, all carbon stock held as standing timber is counted as being released into the atmosphere by clear-cutting (Hiroshima and Nakajima 2006). Therefore, the larger clear-cutting area (Fig. 5) under Scenario 1 decreased the carbon stock dramatically.

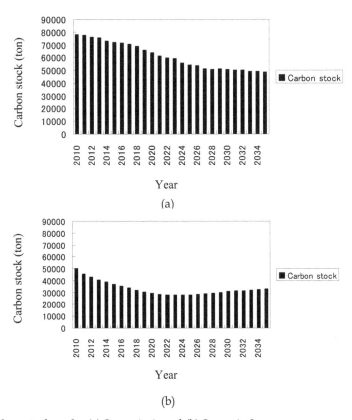

Fig. 7. The carbon stock under (a) Scenario 1, and (b) Scenario 2.

3.3 Subsidy

Figure 8 shows how subsidies vary depending on the scenario. Under Scenario 1, the maximum and minimum subsidies were 32.9 million yen (M¥) in 2017 and 6.8 M¥ in2010; the maximum and minimum silviculture subsidies were 30.0 M¥ in 2017 and 2.4 M¥ in 2010; and the maximum and minimum thinning subsidies were 4.4 M¥ in 2010 and 2.1 M¥ in 2035. Under Scenario 1 the silviculture subsidy was generally larger than the thinning subsidy, with the percentages ranging from 35 % and 65 % in 2010 to 92 % and 8 % in 2016.

After 2017, the subsidies decreased by up to 38.2 % of their maximum value due to a decrease in the harvesting area (Fig. 5a) for clear-cutting. The subsidy in 2035 was 184.6 % of the subsidy in 2010. Under Scenario 2 the maximum and minimum subsidies were 24.9 M¥ in 2017 and 1.0 M¥ in 2010; the maximum and minimum silviculture subsidies were 19.1 M¥ in 2034 and 2.4 M¥ in 2010; and the maximum and minimum thinning subsidies were 7.6 M¥ in 2010 and 4.6 M¥ in 2035. Under Scenario 2 the thinning subsidy in the initial stage of the prediction period was larger than the silviculture subsidy with percentages of silviculture and thinning subsidies ranging from 24 % and 76 % in 2010 to 80 % and 20 % in 2035.

Subsidies increased by up to 248.9 % of their minimum value over the period of simulated predictions due to an increase in the total harvesting area (Fig. 5b). The total subsidy under Scenario 1 is larger than that under Scenario 2 between 2012 and 2021.

A comparison of the two scenarios shows the silviculture subsidy in Scenario 1 of the initial stage under the prediction period to be clearly larger than that of Scenario 2, with differences ranging between 0 and 14.9 M¥. After 2012, the difference of silviculture subsidy decreased by up to 0.2 % of the maximum difference, while the thinning subsidy was clearly larger under Scenario 2 than Scenario 1, with differences ranging between 2.5 and 4.2 M¥.

3.4 Forestry profits

Figure 9 shows the forestry profits under the two scenarios. Under Scenario 1 the maximum and minimum forestry profits were 44.2 M¥ in 2010 and 1.8 M¥ in 2029. After 2011, the forestry profits decreased by up to 4.0 % of their maximum values due to a decrease of harvesting area (Fig. 5a) for clear-cutting.

Under Scenario 2 the maximum and minimum forestry profits were 19.6M¥ in 2011 and 12.2 M¥ in 2035. Between 2010 and 2035 forestry profits decreased by up to 62.4 % of their minimum values due to the increased harvesting area (Fig. 5b). Although the total forestry profits under Scenario 1 are larger than under Scenario 2 up to 2012, the pattern was reversed in 2013.

A comparison of the two scenarios shows the forestry profits under Scenario 1 before 2013 to be larger than under Scenario 2, with differences ranging between 3.4 M¥ and 24.9 M¥.

3.5 Labor requirements

Figure 10 shows the labor requirements under the different scenarios. Under Scenario 1 the maximum and minimum labor requirements were 4647 workers in 2012 and 1830 workers in 2011; the maximum and minimum number of required silviculture workers were 2977 in 2011 and 87 in 2011; the maximum and minimum number of workers for stand thinning were 799 in 2010 and 342 in 2035; and the maximum and minimum number of forest workers for clear-cutting were 1627 in 2010 and 186 in 2034. Under Scenario 1 the labor requirements for clear-cutting and silviculture were generally larger than those for thinning, with the ratio of the proportion of total labor required for clear–cutting to the proportion of the total labor required for thinning ranging from 58 % and 37 % in 2011 to 18 % and 13 % in 2019.

After 2012, the labor requirements decreased by up to 30.5 % of their maximum value. The overall decrease was due to a decrease in the harvesting area (Fig. 5a) for clear-cutting.

Under Scenario 2 the maximum and minimum labor requirements were 3272 personnel in 2030 and 2014 in 2011; the maximum and minimum numbers of workers required in silviculture were 1663 in 2032 and 87 in 2011; the maximum and minimum numbers of people involved in thinning were 1450 in 2010 and 829 in 2035; and the maximum and minimum numbers of clear-cutting forest workers were 661 in 2031 and 498 in 2010. Under Scenario 2 the labor required for clear-cutting and silviculture was generally larger than was required for thinning, with percentages of clear-cutting labor to thinning labor ranging from 25 % and 71 % in 2011 to 20 % and 27 % in 2034. Labor requirements increased by up to 162.5 % of the minimum value between 2011 and 2030, the increase being due to the increase in harvesting area (Fig. 5b).

Labor requirements increased by up to 162.5 % of the minimum value between 2011 and 2030, the increase being due to the increase in harvesting area (Fig. 5b).

A comparison of the two scenarios clearly shows that silviculture requires more workers under Scenario 1 than under Scenario 2 in the initial stage of the prediction period with differences ranging between 0 and 2039 personnel. After 2013, the difference in labor requirements for silviculture decreased by up to 1.0 % of the maximum value. In contrast, the labor required for thinning was greater under Scenario 2 than under Scenario 1, with differences ranging between 486 and 857 personnel. These results suggest that the differences in labor requirements under Scenarios 1 and 2 were mainly associated with silviculture practices and thinning, respectively.

Because the estimated subsidies, forestry profits, carbon stocks, and labor requirements are affected by fluctuations in the stand age distribution and the stand condition over time, the observed pattern of increase was not monotonic.

Our approach enables the effects of different carbon price scenarios on forestry to be calculated. Although timber production is the basic function of forests, their role in storing carbon stock also holds a high position in the public mind, especially during the first commitment period of the Kyoto Protocol. Figures 6 and 9 enable us to consider the influence of forest management under different carbon price on both of these factors. In addition, the simulation results for subsidies and labor requirements can be considered as important practical issues for forest management. Subsidies (Fig. 8) and labor requirements (Fig. 10) under the two scenarios were thus mainly allocated to clear-cutting and thinning (Fig. 5) under Scenarios 1 and 2, respectively. These results suggest that if the clear-cutting area were to decrease (Fig. 5a), the required subsidy (Fig. 8a) and labor (Fig. 10a) would not decrease immediately, because weeding continues to be required for 5 years after planting in the clear-cutting area.

Previous studies have analyzed useful variables and estimated parameters for several econometric models including the probit model (Dennis 1990; Pattanayak et al. 2003) and the logistic regression model (Royer 1987; Zhang and Pearse 1997), which can be used to predict the effects of forestry policies and subsidy systems. Other previous studies (e.g. Lewis and Plantinga 2007; Kurttila et al. 2006; Bolkesjø and Baardsen, 2002) have created models to estimate the effects of different amounts of subsidy. The models used herein mainly made use of established statistical techniques.

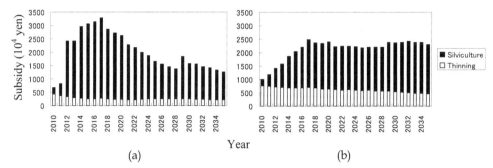

(a)   (b)

Fig. 8. The silviculture and thinning subsidy under (a) Scenario 1, and (b) Scenario 2. White and black blocks show the thinning and silviculture subsidy, respectively.

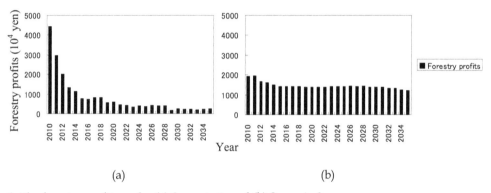

(a)   (b)

9. The forestry profits under (a) Scenario 1, and (b) Scenario 2.

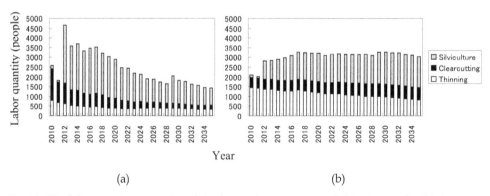

(a)   (b)

Fig. 10. The labor requirements for silviculture, clear-cutting and thinning under (a) Scenario 1, and (b) Scenario 2.

White, black and gray blocks show the thinning, clear-cutting and silviculture labor requirements, respectively.

If policy makers wish to apply these models to other geographical areas, different values for the statistical parameters may be required. We made use of a number of simulations developed and applied to Japan at the national level (Nakajima et al., 2010), so the current models are applicable throughout Japan without the need for new estimates of the parameters. Compared with other studies using similar statistical modeling approaches (Dennis 1990; Pattanayak et al. 2003; Royer 1987; Zhang and Pearse 1997; Lewis and Plantinga 2007; Kurttila et al. 2006; Bolkesjø and Baardsen, 2002), our work appears to be more broadly applicable. Although there may be dramatic changes in carbon and timber prices in the future, our approach should enable us to predict the effect of carbon price scenarios on forest resources and timber production in Japanese forest plantations.

For instance, in the present study, under Scenario 1 it is feasible to increase timber production during the early period of our predicted output (Fig. 6). However, Scenario 2 is a better option if the forests' function of holding carbon stock is the more pressing and stronger requirement (Fig. 9). The most suitable scenario could be selected by considering practical issues based on labor requirements and subsidies (Figs 8 and 10).

As explained in the introduction, Scenario 2 focuses on expanding the thinning area and restricting the clear-cutting area and so supports long-rotation silviculture as a means of increasing the carbon stock as required under the Kyoto Protocol. A comparison of the simulation results of Scenarios 1 and 2 shows that maintaining the carbon stock is more feasible under Scenario 2 (Fig. 7). Because a larger amount of subsidy is available for silviculture (Fig. 8a) following regenerations in the larger clear-cutting area (Fig. 5a). However, if the production of a large amount of timber is not an immediate requirement, Scenario 2 can be the better alternative with a lower subsidy budget. Notwithstanding this, in terms of the efficient use of the timber resource, such a choice might be irrational under some circumstances because of the possibility that some profitable stands might then be forced to avoid clear-cutting in order to produce larger timber.

These simulations can help policy makers and forestry practitioners propose policy changes that would not only enhance timber production, but also fulfill carbon stock obligations pledged under the Kyoto Protocol. Because there was no real and practical system for trading carbon credits at that time, Calish et al. (1978) did not consider the accumulation of carbon credits to be a management objective. The current study clarifies the effect of the Japanese carbon credit trading system on future forest resources. Sakata (2005), similarly, examined the effects of the carbon market on forestry profits in the USA. At the study site selected by Sakata (2005), unlike our site, pulp wood was a second commercially viable product, along with timber. Therefore, the current study shows the effect of the carbon market on forestry profits associated with timber but not pulp wood production.

Planted forests in the present study was conducted are highly productive of timber, especially from the main tree species (*Cryptomeria japonica*). Because this species is very broadly distributed (Fukuda et al. 2003), the simulations described here, which are based on real data, could also be applied to planted forests in other regions. In other words, *Cryptomeria japonica*, which is the target tree species in this study, is the most common tree species in Japanese planted forests (Forestry agency, 2007), so the work is applicable to other parts of Japan. In addition, the growth prediction system used in this study has been applied to the main tree species that grow throughout Japan (Nakajima et al., 2010; 2011a), so this methodology could also be applied to other areas of Japan. Sakata (2005) estimated the effect of the carbon market on the forestry profits based on standard silvicultural practices and costs over a large area including the southeastern United States. Although we considered a standard silvicultural system and costs (Nakajima *et al.*, 2011a) in the present study, we made use of real age distribution and site index data for the study site, which is representative of much of the Japanese planted forest area (The Tokyo University Forests, 2006). We, thus, consider our results to be generally applicable across Japan.

Basically, the cycle for forest management depends on the management objective. Although in the present study, we assumed certain values in order to predict the effect of the real Japanese carbon trading system on timber production and carbon stock, certain socio-economic conditions that are represented by model parameters, could change. However, because the discount rate is the interest rate used to determine the present value of future cash flows (Eatwell et al. 1987; Winton JR. 1951), it is defined relative to a value that society considers to be reasonable. Although a previous study (van Kooten et al. 1995) has stated that, in general, the higher the discount rate, the shorter the rotation period, it is difficult to predict accurately not only the future timber price but also the discount rate as it might be affected by changing socio-economic conditions. Thus, it would be better to improve forest management plans by inputting into the simulation model parameter values that reflect the current socio-economic conditions, and changes in those socio-economic conditions, including discount rates and timber prices that might prevail in the future. Forest management plans could then be simulated by considering, not only socio-economic conditions, but also forest resource productivity and the age distribution of stands derived from forest inventory data.

In the present paper we have described an approach that is designed to increase information concerning objective economic and environmental outcomes of forest management such as timber volume (Fig. 6), forestry profits (Fig. 9) and carbon stock (Fig.7), budgets, operability and subsidies (Fig. 8), labor requirements (Fig. 10). Thus, policy makers could use the information from the simulations designed to understand the influence of different carbon price scenarios on local forestry, to select appropriate plans that would meet their management goals. Other simulation results could be used to decide what information should be taken into consideration when deciding whether or not the benefits of a particular management action would justify the costs of its implementation.

Although there are always uncertainties concerning the future state of socio-economic conditions, the present simulation results can at least provide information about any future tendency of estimated values to change over the prediction period in response to the carbon price scenario currently being implemented under the present socio-economic conditions. However, because estimates are prone to errors derived from a dramatic change in the socio-economic conditions that pertain to forestry, such as timber price, carbon price and discount rate, it is important that the actual forest area continues to be monitored in order to check the accuracy of simulations designed to predict future state of forestry. Although our assumptions concerning socio-economic conditions and forest resources were necessarily relatively simple for the preliminary simulation conducted for the present study, as were the patterns of the different subsidy system scenarios, any uncertainty derived from the future changes in socio-economic conditions should be monitored during the management of regional forest resources.

The next challenge is to test the uncertainty of the simulation by monitoring the study area, and to apply the simulation to other forest regions. Depending on the degree of uncertainty and the wider applicability of the simulation, it may be possible to analyze the feasibility of different management strategies and the efficiencies of different subsidy systems according to different regional forest resources, variations in local socio–economic conditions, and diverse forest management goals.

4. Acknowledgments

We thank the staff of the Tokyo University Forest in Chiba for their valuable assistance in collecting the data set. I thank Dr. Norihiko Shiraishi and Dr. Satoshi Tatsuhara who provided helpful comments to improve this paper. This study was supported in part by a research fellowship from the Ministry of Land, Infrastructure, Transport and Tourism.

5. References

Amano, M., 2008a. "Climate Change and Forest Resources Management", Strategy to Combat Climate Change and the Pacific, IGES, 65-69.

Amano, M., 2008b. Expectation of LiDAR on forest measurement in Kyoto Protocol. Journal of Forest Planning. 13, 275-278.

Amano, M., Tsukada, N., 2006. "Promotion of Sustainable Forest Management under Climate Change Regime", Second Informal Dialogue on the Role of Land Use, Land Use Change and Forestry in the Climate Change Response, Ministry of Environment, Spain, 161-171.

Backe´ us S, Wikstro¨m P, La¨ma° s, T. 2005. A model for regional analysis of carbon sequestration and timber production. Forest Ecology and Management. 216, 28-40. ISSN: 0378-1127

Bolkesjø TF, Baardsen S. 2002. Roundwood supply in Norway: micro-level analysis of self-employed forest owners. Forest Policy and Economics. 4, 55-64. ISSN: 1389-9341

Calish, S.,Fight, RD., Teeguarden, DE.. 1978. How Do Nontimber Values Affect Douglas-Fir Rotations? Journal of Forestry. 76, 217-221.

Campbell, HF., Jennings, SM. 2004.Non-timber values and the optimal forest rotation: An application to the southern forest of Tasmania Source. ECONOMIC RECORD. 80, 387-393.

Dennis, D. 1990. A probit analysis of the harvest decision using pooled time-series and cross-sectional data. Journal of Environmental Economics and Management. 18, 176-187. ISSN: 0095-0696

Eatwell, J., Milgate, M., Newman, P. 1987. The new Palgrave: a dictionary of economics. London: Macmillan.

Ehman, JL., Fan, WH., Randolph, JC., Southworth, J., Welch, NT. 2002. An integrated GIS and modeling approach for assessing the transient response of forests of the southern Great Lakes region to a doubled CO_2 climate. Forest Ecology and Management. 155 (1-3), 237-255. ISSN: 0378-1127

Environmental Ministry. 2009. Japan Verified Emissions Reduction (J-VER) system guideline. Tokyo: Environmental Ministry, 50pp.

Foley, TG., Richter, DD., Galik, CS. 2009. Extending rotation age for carbon sequestration: A cross-protocol comparison of North American forest offsets. Forest Ecology and Management. 259 (2), 201-209. ISSN: 0378-1127

Forestry Agency, 2007. Annual Report on Trends of Forest and Forestry — Fiscal Year 2006 (in Japanese). Tokyo: Japan Forestry Association, 164pp.

Fukuda, M., Iehara, T., Matsumoto, M. 2003. Carbon stock estimates for sugi and hinoki forests in Japan. Forest Ecology and Management. 184, 1–16. ISSN: 0378-1127

Galik, CS., Jackson, RB. 2009. Risks to forest carbon offset projects in a changing climate. Forest Ecology and Management. 257 (11), 2209-2216. ISSN: 0378-1127

Han, K., Youn, Y.-C. 2009.The feasibility of carbon incentives to private forest management in Korea. Climatic Change. 94, 157-168. ISSN: 0165-0009

Hennigar, CR., MacLean, DA., Amos-Binks, LJ. 2008. A novel approach to optimize management strategies for carbon stored in both forests and wood products. Forest Ecology and Management. 256, 786-797. ISSN: 0378-1127

Hirata, Y., Furuya, N., Suzuki, M., Yamamoto, H., 2009. Airborne laser scanning in forest management: individual tree identification and laser pulse penetration in a stand with different levels of thinning. Forest Ecology and Management. 258, 752–760. ISSN: 0378-1127

Hiroshima, T. 2004. Strategy for implementing silvicultural practices in Japanese plantation forests to meet a carbon sequestration goal. Journal of Forest Research. 9, 141–146. ISSN: 1352-2310

Hiroshima, T., Nakajima, T. 2006. Estimation of sequestered carbon in Article-3.4 private planted forests in the first commitment period in Japan. Journal of Forest Research. 11, 427-437. ISSN: 1352-2310

Hiroshima, T., Nakajima, T. 2009. Extracting old-growth planted stands suitable for clear cutting and reforestation using GIS. Mtg. Kanto Br. Jpn. For. Soc. 60, 43–46. (in Japanese)

Houghton, JT. Meira, F. LG., Lim, B., Treanton, K., Mamaty, I., Bonduki, Y., Griggs, DJ., Callander, BA. 1997. Revised 1996 IPCC guidelines for national greenhouse gas inventories, reporting instructions. IPCC/OECD/IEA, Bracknell, 128pp.

Im, EH., Adams, DM., Latta, G.S. 2007. Potential impacts of carbon taxes on carbon flux in western Oregon private forests. Forest Policy and Economics. 9 (8), 1006-1017. ISSN: 0378-1127

IPCC. 2000. Land Use, Land-use Change and Forestry. Special report of the Intergovernmental Panel on Climate Change. Cambridge: Cambridge University Press, 377pp .

IPCC. 2007. Climate Change 2007. Mitigation. Contribution of Working Group III to the Fourth Assessment Report of the Intergovernmental Panel on Climate Change. In: Metz, B., Davidson, O.R., Bosch, P.R., Dave, R., Meyer, L.A. (Eds.). Cambridge, Cambridge University Press: 852pp.

Kaipainen, T., Liski, J., Pussinen, A., Karjalainen, T. 2004. Managing carbon sinks by changing rotation length in European forests. Environmental Science & Policy. 7 (3), 205-219. ISSN: 1462-9011

Komaki, T. 2006. The future forestry policy for private stands. The forestry economic research. 52, 1-9. (in Japanese with English summary)

Kurttila, M., Pykalainen, J., Leskinen, P. 2006. Defining the forest landowner's utility-loss compensative subsidy level for a biodiversity object. European Journal of Forest Research. 125, 67-78. ISSN: 1612-4669

Lewis, DJ., Plantinga, AJ. 2007. Policies for habitat fragmentation: combining econometrics with GIS-based landscape simulations. Land Economics. 83, 109-127. ISSN: 0023-7639

Liski, J., Pussinen, A., Pingoud, K., Makipaa, R., Karjalainen, T., 2001. Which rotation length is favourable to carbon sequestration? Canadian Journal of Forest Research. 31, 2004-2013. ISSN: 0045-5067

Matala, J., Karkkainen, L., Harkonen, K., et al. 2009. Carbon sequestration in the growing stock of trees in Finland under different cutting and climate scenarios. European Journal of Forest Research. 128, 493-504. ISSN: 1612-4669

Nakajima, T., Hiroshima, T., Shiraishi, N. 2007a. An analysis of managed private forest focusing on the expansion of lands under the article 3.4 of the Kyoto Protocol. Journal of Japanese Forestry Society. 89, 167-173. (in Japanese with English summary)

Nakajima, T., Hiroshima, T., Shiraishi, N. 2007b. An analysis about local silvicultural practices and subsidy system in a region level. Journal of Forest Planning. 41(2), 179-186. (in Japanese with English summary)

Case Study of the Effects of the Japanese Verified Emissions Reduction (J-VER) System on Joint Forest
Production of Timber and Carbon Sequestration
107

Nakajima, T., Kanomata, H., Matsumoto, M., Tatsuhara, S., Shiraishi, N. 2009a. The application of "Wood Max" for total optimization of forestry profits based on joint implementation silvicultural practices. Kyushu Journal of Forest Research. 62, 176-180.

Nakajima, T., Lee, J.S., Kawaguchi, T., Tatsuhara, S., Shiraishi, N. 2009b. Risk assessment of wind disturbance in Japanese mountain forests. Ecoscience. 16:58-65. ISSN: 1195-6860

Nakajima, T., Matsumoto, M., Tatsuhara, S. 2009c. Development and application of an algorithm to estimate and maximize stumpage price based on timber market and stand conditions. Journal of Forest Planning. 15, 21-27.

Nakajima, T., Matsumoto, M., Sasakawa, H., Ishibashi, S., Tatsuhara, S. 2010. Estimation of growth parameters within the Local Yield table Construction System for planted forests throughout Japan. Journal of Forest Planning. 15, 99-108.

Nakajima,T., Kanomata, H., Matsumoto, M., Tatsuhara, S., Shiraishi, N. 2011a. Cost-effectiveness analysis of subsidy schemes for industrial timber development and carbon sequestration in Japanese forest plantations. Journal of Forestry Research 22, 1-12. ISSN: 1007-662X(print)

Nakajima,T., Kanomata, H., Matsumoto, M., Tatsuhara, S., Shiraishi, N. 2011b. Simulation depending on subsidy scenarios for carbon stock and industrial timber development. FORMATH 11, 143-168.

Nakajima, T., Matsumoto, M., Sakata, K., Tatsuhara, S. 2011c. Effects of the Japanese carbon offset system on optimum rotation periods and forestry profits. International Journal of Ecological Economics & Statistics. 21, 1-18. ISSN: 0973-7537

Nakajima, T., Matsumoto, M., Shiraishi, N. 2011d. Modeling diameter growth and self-thinning in planted Sugi (*Cryptomeria japonica*) stands. The Open Forest Science Journal. 4, 49-56. ISSN: 1874- 3986

Ohmura, K., Sawada, H., Oohata, S. 2004. Growth records on the artificial forest permanent plots in the Tokyo University Forest in Chichibu. Miscellaneous Information, Tokyo University Forests. 43, 1-192. (in Japanese)

Ohtsuka, T., Sakura, T., ohsawa, M. 1993. Early herbaceous succession along a topographical gradient on forest clear-felling sites in mountainous terrain central Japan. Ecological Research: 8: 329-340. ISSN: 0912-3814

Pohjola, J., Valsta, L. 2007. Carbon credits and management of Scots pine and Norway spruce stands in Finland. Forest Policy and Economics. 9, 789-798. ISSN: 1389-9341

Pattanayak, SK., Murray, BC., Abt, R. 2002. How joint in joint forest production: an econometric analysis of timber supply conditional on endogenous amenity values. Forest Science 48 (3), 479-491. ISSN: 0015-749X

Pienaar LV, Turnbull KJ. 1978. The Chapman-Richards generalization of Von Bertalanffy's growth model for basal area growth and yield in even-aged stands. Forest Science 19, 2–22. ISSN: 0015-749X

Pukkala, T. 2002. Multi-objective forest planning. Boston: Kluwer Academic, 207pp.

Raymer, A., Gobakken, T., Solberg, B., Hoen, H., Bergseng, E. 2009. A forest optimisation model including carbon flows: Application to a forest in Norway. Forest Ecology and Management. 258, 579-589. ISSN: 0378-1127

Ravindranath, NH., Somashekhar, BS. 1995. Potential and economics of forestry options for carbon sequestration in India. BIOMASS & BIOENERGY. 8 (5), 323-336. ISSN: 0961-9534

Richards, K., Stokes, C., 1994. Regional studies of carbon sequestration.US Department of Energy DE-AC76RLO 1830, Washington DC.

Royer, J. 1987. Determinants of reforestation behavior among southern landowners. Forest
    Science. 33 (3), 654-667. ISSN: 0015-749X

Sakata, K. 2005. Carbon dioxide price from which an afforestation interest rate becomes
    maximum in each cutting age: loblolly pine in southeast Georgia, USA. Journal of
    Forest Research. 10 (5), 385-390. ISSN: 1341-6979

Sakura, T., 1999. Investigation regarding the vegetation and dynamics of weeds under the
    sugi (*Cryptomeria japonica*) planted forests. Doctor thesis of the University of Tokyo.

Science Council. 2008. The policy to progress research development of global environmental
    science technology. Tokyo: Japanese Ministry of Education, Culture, Sports, Science
    and Technology, 40pp. (in Japanese)

Seidl, R., Rammer, W., Jager, D., Currie, WS., Lexer, MJ.2007. Assessing trade-offs between
    carbon sequestration and timber production within a framework of multi-purpose
    forestry in Austria. Forest Ecology and Management. 248 (1-2), 64-79. ISSN: 0378-1127

Shiraishi, N. 1986. Study on the growth prediction of even-aged stands. Bulletin of Tokyo
    University Forest. 75,199-256. (in Japanese with English summary)

Schroeder, P., 1992. Carbon storage potential of short rotation tropical tree plantations.
    Forest Ecology and Management. 50, 31-41. ISSN: 0378-1127

Sivrikaya, F., Keles, S., Cakir, G.. 2007. Spatial distribution and temporal change of carbon
    storage in timber biomass of two different forest management units. Environmental
    Monitoring and Assessment. 132 (1-3) , 429-438. ISSN: 0073-4721

Thompson, MP., Adams, D., Sessions, J. 2009. Radiative forcing and the optimal rotation
    age: Ecological Economics. 68 (10),2713-2720. ISSN: 0308-597X

Tokyo Stock Exchange. 2007. Annual statistics of stock exchange. CD-ROM. Tokyo Stock
    Exchange. Tokyo.

The Tokyo University Forests, 2006. Annual Report of Meteorological Observations in the
    Universities Forests (in Japanese). The University of Tokyo (January 2004–December
    2004). Miscellaneous Information the Tokyo University Forest 45, pp.271–295.

Tsuyuki, S., Nakajima, T., Tatsuhara, T., Shiraishi, N. 2011. Analysis of natural wind
    disturbance regimes resulting from typhoons using numerical airflow modelling
    and GIS:A case study in Sugi (*Cryptomeria japonica*) and Hinoki (*Chamaecyparis
    obtusa*) plantation forests. FORMATH. 10: 87-103.

UNFCCC. 1998. Report of the Conference of the Parties on its third session, held at Kyoto,
    from 1 to 11 December 1997. Addendum. Part two: Action taken by the Conference
    of the Parties at its third session. United Nations Office at Geneva: Geneva, 60pp.

UNFCCC. 2002. Report of the Conference of the Parties on its seventh session, held at Marrakesh
    from 29 October to 10 November 2001. Addendum. Part two: Action taken by the
    Conference of the Parties. Volume I. United Nations Office at Geneva: Geneva, 69pp.

van Kooten, GC., Binkley, CS., Delcourt, G. 1995. Effect of carbon taxes and subsidies on
    optimal forest rotation age and supply of carbon services. American Journal of
    Agricultural Economics. 77, 365-374. ISSN: 0002-1962

Winton, JR. 1951. A dictionary of economic terms: for the use of newspaper readers and
    students. London: Routledge & K. Paul.

Zhang, D., Pearse, P. 1997. The influence of the form of tenure on reforestation in British
    Columbia. Forest Ecology and Management. 98, 239- 250. ISSN: 0378-1127

Section 3

Forest Health

6

Cambial Cell Production and Structure of Xylem and Phloem as an Indicator of Tree Vitality: A Review

Jožica Gričar
Department of Yield and Silviculture, Slovenian Forestry Institute
Slovenia

1. Introduction

One third of Europe's land surface is covered by forests, with important economic and social value. They constitute the most natural ecosystems of the continent. Natural biotic and abiotic disturbances affect their structure and composition. Sustainable forest management and environmental policies rely on the sound scientific resource provided by long-term, large scale and intensive monitoring of forests. Long-living trees and ecosystems are suitable for studying the impact of human factors as opposed to the effects of natural system variability. Forest monitoring helps to improve our knowledge of the state of forests and to quantify changes that are taking place within forests and related ecosystems. Information about forest ecosystem functions and processes is, however, also necessary to gain an understanding of the causes underlying such changes and, subsequently, to model the future effects of natural and anthropogenic stress factors on our forests and understand the adaptation potential of forests to the effects of environmental change and air pollution (UN-ECE, 2008).

The vitality of trees is one of the most important indicators of forest condition and can be characterized by different parameters, such as assessment of the crown condition, tree growth etc. (UN-ECE, 2008). However, the latest studies show that cambium activity and increments of its products – secondary phloem and secondary xylem (wood) – reflect the vitality of trees (Gričar et al., 2009). In the different vitality of silver firs (*Abies alba* Mill.), it has been demonstrated that data on phloem increment structure, the relationship between the phloem and wood increment and the number of cambial cells in the dormant state are very useful in the assessment of the health condition of trees.

This review focuses in particular on presenting the potential of structure and width of xylem, phloem and cambium in the case silver fir and pedunculate oak, as indicators of the vitality status of trees. Forest monitoring and indicators of tree vitality status will be briefly summarized. Growth ring patterns have proved to be an appropriate tool for quantifying the response of a forest stand to changing environmental factors, so wood formation processes that determine the structure of wood and its quality will be described in more detail. Finally, tree vitality has a major influence on wood quality. Two examples, silver fir and pedunculate oak, will therefore be demonstrated.

2. Forest monitoring

Only healthy and vital forests can serve multiple ecological, social and productive roles, as understood by the modern world. To be able to acquire a reliable assessment of the state and changes in forest ecosystems and at least partly to explain and understand the most important processes occurring in them, forest monitoring programmes have been implemented worldwide (e.g., International Co-operative Programme on Assessment and Monitoring of Air Pollution Effects on Forests – ICP Forests, Acid Deposition Monitoring Network in East Asia – EANET, Forest Health Monitoring – FHM). Monitoring is essential in order to obtain information about the condition of natural resources, their development over time and space, and to study their relationships with biotic/abiotic factors (Ferretti, 2009). Environmental and nature management, namely, cannot operate effectively without reliable information on changes in the environment and on the causes of those changes. There is therefore considerable concern in the scientific community about the ability of monitoring programmes to provide the desired information (Legg & Nagy, 2006; Vos et al., 2000). Some researchers believe that many operational monitoring programs are not very effective or useful for decision-making (Vos et al., 2000). The main reason for this is poor confidence in the quality of the data, with the most typical questions raised about the statistical basis of sampling design, the reliability and comparability of data and data management (e.g., Ferretti, 2009; Legg & Nagy, 2006; Vos et al., 2000). The results of inadequate monitoring are misleading in terms of their information quality and are dangerous because they create the illusion that something useful has been done (Legg & Nagy, 2006).

Indicators of tree health and vitality need to be accurate and reliable, but also cheap and easy to use (Martín-García et al., 2009). The quality of monitoring is thus defined by its ability to provide data that allow estimates of the status of the target resource with the defined precision level, permit the detection of change with the defined power, and are comparable through space and time. Despite considerable work on data quality control, parts of the monitoring process are still poorly covered by quality assurance and have revealed weaknesses in design and implementation. Steps towards a more comprehensive quality assurance approach have currently been undertaken (Ferretti, 2009).

3. ICP Forests

In Europe, the International Co-operative Programme on the Assessment and Monitoring of Air Pollution Effects on Forests (ICP Forests) was established in 1985. The system combines an inventory approach with intensive monitoring. It provides quality assured and representative data on forest ecosystem health and vitality and helps to detect responses of forest ecosystems to the changing environment. Air pollution effects are the particular focus of the programme. ICP Forests uses two complementary monitoring approaches on the European level. Representative monitoring (Level I) is based on around 6,000 permanent observation plots and provides an annual overview of forest condition on the European level. Intensive monitoring (Level II) on around 500 sites provides insight into factors affecting the condition of forest ecosystems and into the effects and interactions of different stress factors. These plots are located in forests that represent the most important forest ecosystems of the continent. The programme provides an early warning system for the impact of environmental stress factors on forest ecosystem health and vitality. Although forest species have responded to environmental changes throughout their evolutionary history, a primary concern in relation to wild species and their ecosystems is the rapid rate of human induced changes (UN-ECE, 2008).

4. Vitality of forest and trees

Though the vitality of trees is one of the most important indicators of forest condition, forest health cannot be assessed solely on the basis of tree condition, since forest consists of more than trees (Innes, 1993). Tree vigour is best restricted as a term to the growth of trees in relation to a hypothetical optimum, whereas tree health is defined in the pathological sense as the incidence of biotic and abiotic factors affecting the tree within a forest. Tree condition is less specific, referring to the overall appearance of trees within the forest. The health of a tree can be evaluated by such indicators as crown condition, growth rate and external signs of disease-causing agents (Kolb et al., 1994).

Tree vitality cannot be measured directly, only through several indicators, such as assessment of the crown condition, growth of bud, stem (radial or height) and root systems, measurements of cambial electrical resistance or the size and shape of needles etc. (Dobbertin, 2005). Shigo (1986) defines vigour as the capacity of a tree to resist strain. It determines the potential strength against any threats to survival. It is genetically derived and cannot therefore be changed. Tree vitality, on the other hand, is the dynamic ability of a tree to grow under the conditions present. It is important to assess the influence of external stress, since resistance to stress is an important criterion in the vitality concept (Dobbertin, 2005). Larcher (2003) defines stress as a significant deviation from the condition optimal for life. Vitality becomes weaker as stress persists. At a certain point, the capacity of a tree to overcome further stress or to survive diminishes, i.e., vitality decreases. Irreversible damage or tree death can occur. The hypothetical optimal tree vitality is not known; only the minimum vitality (i.e., tree death) can be identified (Dobbertin, 2005).

The consequences of tree death, in terms of effects on other ecosystem components and processes, depend on many variables, including the species, mortality agent, position, spatial pattern (dispersed or aggregated) and numbers that have died. Tree death is an important indicator of ecosystem health and can assist recognition of stresses caused by pollutants, such as acid rain and ozone. However, the value of tree death as an indicator of anthropogenic disturbance depends on a thorough understanding of the patterns of tree death under natural conditions. At the present time, adequate understanding of this is woefully lacking. Tree death also demonstrates some principles of ecological processes: the importance of defining the spatial and temporal context of a study, the importance of stochastic processes, and the fact that most ecological processes are driven by multiple mechanisms and that the relative importance of these mechanisms changes over time and space, and the importance of the natural histories of species and ecosystems. Tree death illustrates that many valid and useful perspectives on a single, presumably simple process exist. Furthermore, it makes clear that we need to give more consideration to the biology of organisms and ecosystems in developing, evaluating and applying theoretical constructs (Franklin et al., 1987).

It is not possible to estimate forest health condition from concepts developed at the individual organism level and simply to apply them on a landscape level (Kolb et al., 1994). In other words, extension of this concept to a complex system, such as a forest, is based on making an analogy between the functioning of an organism and an ecosystem. A dead or dying tree is not healthy. The health of a stand must take into account many more dimensions than the health of a tree. The health of a stand relates to the management objectives for that stand (utilitarian perspective) and to the long-term functioning of the organisms and trophic networks that constitute the stand (ecosystem perspective). Tree mortality in a stand would not indicate an unhealthy condition as long as the rate of mortality was not greater than the capacity for replacement. Stand objectives such as wildlife habitat, soil and water protection and preservation of biodiversity do not require that all trees be healthy. A dead tree is not healthy but it may be part of a healthy stand. The health of a forest ecosystem or landscape is similarly more complex than the health of a stand (Kolb et al., 1994).

5. Indicators of tree vitality status

Biochemical indicators on the plant cell level, such as phytohormones or enzymes, may best reflect the reaction of trees to various environmental stresses (Larcher, 2003). Unfortunately, many such indicators cannot readily be extracted in the field or are very expensive. Several indicators, such as assessment of the crown condition, growth of bud, stem (radial or height) and root system, measurements of cambial electrical resistance or size and shape of needles etc. may instead be used (Dobbertin, 2005).

Crown condition is a major indicator of forest health in Europe. The condition of forest trees in Europe is monitored over large areas by a survey of tree crown transparency and discoloration, which is a fast reacting indicator of numerous natural and anthropogenic factors affecting tree vitality. Crown transparency and discoloration is a valuable indicator of the condition of forest trees. It reflects, among other factors, weather conditions and the occurrence of insects and fungal diseases. Such information is extremely relevant for monitoring the reactions of forest ecosystems to climate change and for ensuring sustainable forest management in the future (UN-ECE, 2008).

Crown condition assessments are commonly used in monitoring programs, since they are quick, easy and cheap to do. However, interpretation of these data may be complicated by the occurrence of strong fructification years in some tree species, when the foliage is reduced (Beck, 2009). In addition, tree crown transparency and discoloration used to be visually estimated by observers from the ground, raising the questions about the subjectivity of human assessment, data quality and comparability across the countries (Mizoue & Dobbertin, 2003). These issues were tried to be solved by combining field and control team assessments, using data from cross-calibration courses to estimate correction factors, using reference photographs or standard sets of two-dimensional silhouettes representing various degrees of foliar density (Frampton et al., 2001; Ghosh & Innes, 1995; Innes et al., 1993; Solberg & Strand, 1999). Nevertheless, since these improvements were sometimes not enough, the researchers started to replace visual ground assessment by digital photo (Martín-García et al., 2009; Mizoue, 2002) or by remote sensing techniques (Coops et al., 2004; Stone et al., 2003)

Moreover, it is not possible simply to conclude tree vitality by crown condition. Growth rates can be considerably reduced while foliage is still inconspicuous (Beck, 2009). In particular, the relation between crown condition and xylem increment in trees has not been satisfactorily explained. In principle, physiological investigations of these relationships through case studies may be useful for improving our understanding.

As summarised by Kozlowsky & Pallardy (1997), the requirements for tree growth are carbon dioxide, water, and minerals for raw materials, light as an energy resource, oxygen and favourable temperature for growth processes. The capacity of photosynthetic processes (i.e., foliar biomass) and competition for resources are constraining factors for tree growth. Tree growth processes can be ranked by order of importance as foliage growth, root growth, bud growth, storage tissue growth, stem growth, growth of defence compounds and reproductive growth (Waring, 1987; Waring et al., 1980). Under stress, photosynthesis is reduced and carbon allocation is altered. Stem growth may be reduced early on, since it is not directly vital to the tree. Comparison with a suitable reference is important for any potential vitality indicator. Depending on the aim of the study, the references used can be the growth of trees presumed to be without stress. The general disadvantage is that no absolute growth reference is available. Some stresses, such as competition, root rot or mistletoe occurrence, affect the tree over extended time periods, whereas other stresses, such as drought or insect defoliation, cause immediate reactions. Annual or inter-annual stem growth assessment is therefore needed in long-term monitoring plots. Tree growth can serve as a vitality indicator if a reference growth or growth trends are available (Dobbertin, 2005). It is noteworthy that not every stress is necessarily negative for trees but can instead induce increased resistance to stress (Kozlowsky & Pallardy, 1997; Larcher, 2003). A short-term stress reaction may therefore not coincide with a long-term change in tree vitality. Growth changes must thus be interpreted on a long-term perspective (Dobbertin, 2005).

6. Growth ring patterns as indicators of tree health

Beck (2009) emphasized dendroecological analysis of tree and stand growth patterns as an appropriate tool for quantifying the response of a forest stand to changing environmental factors. Tree growth parameters, which reflect changing growth conditions year by year, are very important. Such parameters of tree vigour presented as a time series retrospectively enable an insight into the growth history of the stand. Namely, tree-ring analysis can provide information on trees and stand development in the past. The growth of trees and site history of a stand can be reconstructed using tree-ring time series, which contain lots of information on environmental conditions and their impact on the growth of trees. Wood formation is the final result of the complete metabolic balance. It is the share of the balance of matter produced by the foliage, respiration and the higher priorities of allocation to other tree organs (roots, fruits etc.), which has not been consumed elsewhere. This remaining share refers directly to the state of the reserve pool. The amount of new wood formation can therefore be understood as a suitable tree health indicator. In contrast to visually assessed crown condition, tree-ring widths are measured (qualitative) data. Subjective estimation is thus eliminated (Beck, 2009).

Tree growth rates are affected by both pollutants and climatic stress. In view of this complexity, a comprehensive dendroecological analysis of tree and stand growth patterns is considered to be an appropriate tool for quantifying the response of a forest stand to changing environmental factors. The inclusion of dendroecological methods in monitoring programs provides many advantages and new findings. The elaboration and analysis of tree ring networks (chronologies well scattered with respect to space and altitude) are currently seen as successful fields of ecological research. Ongoing depositions and increasing climatic stress urgently require the quantification of the growth response of forests as an indicator of tree and stand vigour (Beck, 2009).

Decreasing growth curves are among the most obvious growth-related characteristics of dying trees, which is not only a species-specific but also a site-specific feature. A combination of growth levels or relative growth and growth trends has been shown to increase the reliability of mortality predictions. Abrupt declines in growth or strongly negative growth trends may indicate a rapid physiological adaptation to changed environmental conditions (Bigler et al., 2004). Reduced xylem increments, as one of the first indicators of decreased tree vitality, are very useful for reconstruction of past tree vitality and evaluation of mortality risk (Bigler & Bugmann, 2004; Bigler et al., 2004). These assessments of individual tree vitality and accurate mortality predictions may be used in forest management to identify and selectively cut low-vitality trees, so as to release the remaining healthy trees (Bigler et al., 2004).

7. The role of cambium in a tree

The growth of trees, leading to an increase in the size and mass of an organism, occurs only in specific areas, so called meristems, and involves cell division, expansion and differentiation. Cell division is an essential part of growth, resulting in an increase in the number of cells. In the expansion phase, cells increase in size. The cytoplasm grows and the vacuole fills with water, which exerts pressure on the cell wall and causes it to expand. In the next step, cells differentiate, or specialize, into various cell types. A tree is composed of various cell types that perform different functions required in a multicellular organism (Berg, 2008).

Meristems consist of actively dividing undifferentiated cells, which retain the capacity for growth through their entire lifespan. Two kinds of meristematic growth occur in trees; primary or extensional and secondary or lateral (Berg, 2008). Primary growth occurs as a result of the activity of apical meristems, which are located at the tips of stems and roots and lead to an increase in a tree's length and development of various tissues: epidermis, cortex, conducting veins, pith and leaves (Fig. 1) (Mauseth, 1988). In woody plants, even in the first year of growth the primary tissues of stems and roots are replaced by secondary tissues formed by the secondary lateral meristems: vascular cambium (in short cambium) and cork cambium (phellogen). The activity of secondary meristem is expressed as radial growth, which allows the increase of the volume of the conducting system and the formation of mechanical and protective tissues (Plomion et al., 2001; Taiz & Zeiger, 2002). Cambium produces secondary vascular tissues, which conduct water and nutrients and provide support. Cork cambium produces protective tissue (periderm), which protects the stem and root from water loss, pathogens, and herbivorous insects (Larson, 1994; Panshin & de Zeeuw, 1980).

The cambium, as an uninterrupted, thin layer ring, lies between the secondary xylem on the inner side and secondary phloem on the outer side, the two tissues it produces (Larson, 1994; Mauseth, 1988). It has been called the "least understood plant meristem", because of the associated technical difficulties when working with trees (Groover, 2005).

The cambium consists of a layer of cells that divide actively, have small radial dimensions and have no intercellular space (Savidge, 2001). It differs from other meristems by two types of highly vacuolised cells: short, rather isodiametric ray cells, from which radial rays are formed, and elongated fusiform or spindle-shaped cells, which form axial elements (Fig. 2). New cambial cells are formed by anticlinal divisions, which ensures an increase in girth of the cambium. The cambium has a decisive role in radial growth and the development of trees, since new vascular tissues of xylem and phloem are formed through periclinal or additive divisions that occur in the tangential plane, by which the diameter of the tree increases. The production of secondary conductive tissues represents approximately 90% of all mitoses (Lachaud et al., 1999; Larson, 1994).

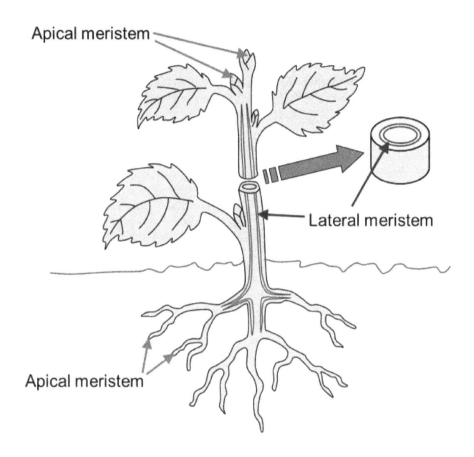

Fig. 1. Location of apical and lateral meristems in a plant.

A characteristic of tree species in the temperate climatic zone is a seasonal alternation of cambial activity and dormant (resting) periods, which is generally related to alternations of cold and hot or rainy and dry seasons (Larcher, 2003). Cambial activity usually starts in spring with cell division and ends in late summer with the completed development of the latest newly formed cells (Fig. 2). At the beginning of cambial activity, the number of cambial cells increases and they start to divide, which is followed by differentiation of derivatives into the adult elements of xylem or phloem. In the process of differentiation, which includes post-cambial cell growth, deposition of the secondary cell wall and – in wood tracheids, fibers and vessels – also lignification and programmed cell death, the cells specialize in order to perform their functions (Fig. 3) (Plomion et al., 2001). The vascular system in trees is very complex, composed of various types of cells, which are differently orientated (Chaffey, 2002).

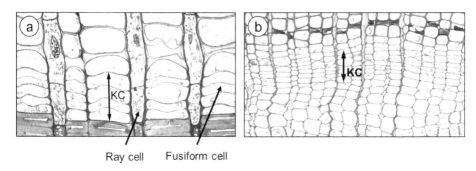

Fig. 2. (a) Dormant and (b) active cambium. KC – cambial cells

1. Cell divison in a cambium
2. Cell expansion
3. Secondary wall formation
4. Lignification
5. Programmed cell death

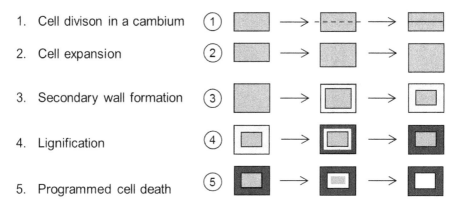

Fig. 3. Schematic illustration of formation of tracheid from cambial cell in conifers.

8. Wood and phloem formation

Xylo- and phloemogenesis are periodic processes driven by a variety of internal and external factors, the influence of which changes during the growing season. Xylem and phloem increments are not predetermined, but are plastic end-products of interactions between the genotype and the environment (Savidge, 2001). The environment determines the physical conditions and the energy for xylo- and phloemogenesis. The external factors affect the onset, end and rate of individual growth processes, which determine the morphology of cells (Wodzicki, 2001). Xylo- and phloemogenesis lead to specialization of cells in terms of their chemical composition, morphological characteristics and function. Cell divisions in the cambium and post-cambial growth determine the width of the annual xylem and phloem increment, and the deposition of the secondary cell wall (and lignification) determines the accumulation of biomass in the walls of the xylem and phloem cells (annual biomass increment) (Fig. 3) (Plomion et al., 2001).

The number of dormant cambial cells depends on several factors; such as tree species, tree age, part of the tree, and tree vigour and vitality. The cambium's cell production, under normal growth conditions, is more intensive on the xylem than on the phloem side. However, under physiologically very demanding conditions, the phloem increment can exceed the xylem one, which may not appear at all in exceptional cases (Larson, 1994; Panshin & de Zeeuw, 1980).

Of all the secondary tissues, xylem and its formation is by far the most investigated, particularly due to its great economic and ecological importance. The width and structure of xylem growth rings is a source of information about past and present factors affecting the development processes in an individual tree (Fritts, 1976; Wimmer, 2002). The width of the xylem increment is closely related to its anatomical structure, which defines the physical and mechanical properties of wood and, consequently, its end-use (Fig. 4).

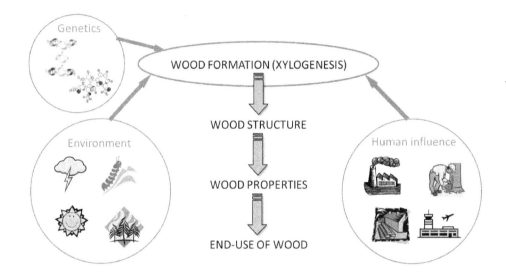

Fig. 4. Xylogenesis is affected by a variety of internal and external factors, which influence wood structure and properties and, consequently, the end-use of wood.

Studies of the seasonal dynamics of phloem growth rings are fewer, which can be partly explained by a lower interest in the commercial use of bark in comparison to the use of timber. In addition, the phloem increment is exposed to relatively fast secondary changes of the tissue, e.g., collapse, sclerification and inflation of axial parenchyma, so only the structure of one or two of the most recent phloem growth rings can be seen clearly. Older non-conducting tissue eventually collapses in a radial direction, deforms and later often also falls off and is thus not suitable for dendrochronological and dendroecological studies (Gričar, 2009).

Nevertheless, the seasonal dynamics of phloem formation is very important in studies of trees' radial growth because cambium is a bi-facial meristem, so studies of cambial activity and wood formation reveal only part of the information on cambial cell productivity during the growth season. Moreover, the processes of wood and phloem formation differ in terms of time and space, and internal and external influences affect the mechanisms of their formation differently. Phloem increment is more stable and less subjected to fluctuations of environmental conditions. Comprehensive studies are therefore vital for investigating the influence of specific climatic factors on the radial growth of trees.

The impact of changing environment will modify the seasonality and rate of growth, which can have an important effect on tree performance and survival and also on wood structure and properties and, consequently, on the end-use of wood.

9. Wood density in relation to ring widths

Wood formation determines the morphology of cells, the structure of the xylem growth ring and thus the wood properties. Xylem rings are composed of early wood and late wood. Early wood cells are formed at the beginning of the growing season and are characterized by a large radial dimension and thin cell walls. The development of late wood cells with small radial dimensions and thick cell walls occurs in summer, resulting in its higher density. In sapwood, the ratio between the density of early wood and late wood is 1: 2.3 in fir and 1: 4.0 in pine (Gorišek, 2009). In ring-porous deciduous tree species this ratio is about 1: 2.5 and in diffuse-porous trees much smaller, e.g., 1: 1.5 in beech (Gorišek, 2009).

Wood is heterogeneous material composed of various types of cells that perform different functions. Consequently, the density of wood is related to the morphological characteristics of the cells. Growth of trees from temperate climate regions is seasonal resulting in the formation of growth rings. At the beginning of the growing season the dominant function appears to be conduction, while in the second part of the season is support. Early wood cells have therefore thinner cell walls and bigger cavities than late wood ones. Hence, the greater is the proportion of late wood the greater are the density and strength. However, wood density and its strength are influenced by the ring width (Fig. 5) (Dinwoodie, 1981). This relationship is relatively complex; in ring-porous tree species, such as oak and ash, increasing ring width results in an increase in the percentage of late wood, which contains most of the fibres and, consequently, the density will increase (Fig. 6). In diffuse-porous tree

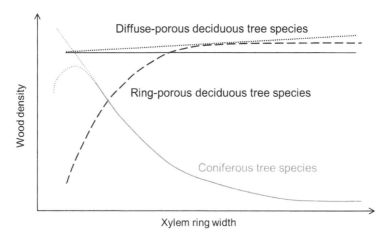

Fig. 5. Impact of xylem ring width on the density of wood in ring-porous and diffuse-porous deciduous trees and conifers.

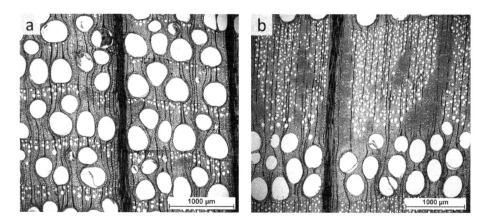

Fig. 6. Wood density is closely related to ring width in ring-porous oak; non-vital (a) and vital oak (b).

species (beech and maple, for example), in which the wood anatomical structure in the xylem ring is relatively homogenous, increasing ring width has almost no effect on wood density. In softwoods, however, increasing ring width results in an increased percentage of low-density early wood and, consequently, a decrease in density (Fig. 7). Exceptionally, softwoods from very cold areas may have narrow rings with low density, because late wood formation is restricted by the short summer period (Dinwoodie, 1981).

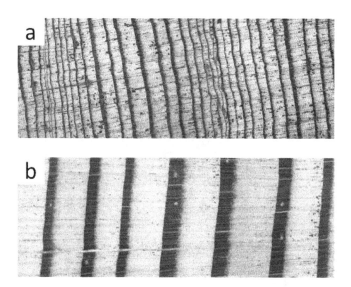

Fig. 7. Widths of xylem rings in European larch, with different proportions of early wood and late wood.

Of the wood properties that affect quality, basic density is one of the most important because it determines its utilization in sawmills, manufacturing factories, cellulose plants and as planks. Several factors influence the variability in basic density: site, climate, geographic location, species, age and silviculture. However, tree vitality also has a major effect on wood quality and properties; the relationship among all this parameters therefore deserves deeper investigation.

10. Relationship among the number of cells in xylem, phloem and dormant cambium in silver fir (*Abies alba* Mill.) trees of different vitality

Silver fir *(Abies alba)* decline has appeared in many European countries, including Slovenia, since about 1500. The exact cause of silver fir decline is still not satisfactorily explained; however, it has been interpreted as a complex disease due to the interaction of several unfavourable factors, such as drought, frost, pollution, competition among trees, soil acidification, inappropriate silvicultural treatments, insects, pathogens etc (e.g., Bauch, 1986; Dobbertin, 2005; Fink, 1986; Schweingruber, 1986; Torelli et al., 1986).

Decline is characterized by reduced cambial production, especially towards the xylem, shorter cambial activity and crown damage, including needle loss and yellowing foliage (Fig. 8, 9) (e.g. Bauch, 1986; Fink, 1986; Innes, 1993; Schmitt et al., 2003; Schweingruber, 1986; Torelli et al., 1999). Reduced wood formation often occurs prior to visual symptoms of crown decline (Torelli et al., 1986, 1999), which highlights the usefulness of assessment of a tree's current mortality risk based on growth patterns and a derived statistical mortality model that clearly identifies trees at high risk of dying (Bigler et al., 2004).

Fig. 8. Crowns of differently vital silver firs.

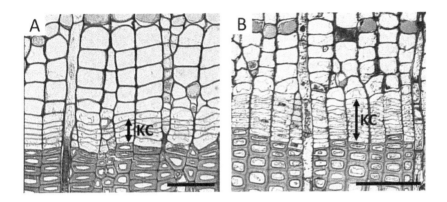

Fig. 9. Narrow cambium in declining silver fir, consisting of four to five cell layers (a) and a wide cambium in healthy silver fir consisting of about ten cell layers. KC – cambial cells, Scale bars = 50 μm (A), 100 μm (B)

Although some papers have been published concerning the anatomical structure and dynamics of secondary phloem formation (e.g., Gričar & Čufar, 2008), the relationship among the number of cells in phloem, xylem and dormant cambium is still poorly understood. In a paper published in 2009, we investigated the anatomical structure of phloem and xylem growth rings, as well as dormant cambium in relation to vitality in 81 adult silver fir trees (*Abies alba* Mill.) (Gričar et al., 2009). Specifically, we investigated the number of cells produced in the current phloem growth ring, xylem growth ring and their ratio, the number of cells in the dormant cambium and the structure of the phloem growth ring, which included characterization of early phloem, late phloem and the presence, absence and continuity of tangential bands of axial parenchyma.

The silver fir (*Abies alba* Mill.) trees were located in an *Abieti-fagetum dinaricum* mixed forest at Ravnik, Slovenia (approx. 45°52′N, 14°16′E, elevation 500-700 m). The studied trees were dominant or co-dominant, with an age of 150–180 years and DBH greater than 50 cm. The trees belonged to a population of 269 mature trees monitored from 1987 to 2007. The health condition of the trees was assessed by determining the crown status index based on progressive needle loss and cambial electrical resistance (CER) (Torelli et al., 1999). Trees were assigned to 3 categories: A – trees with a full crown and productive cambium; B – trees with intermediate characteristics and C – trees with a sparse crown and suppressed cambium (Fig. 8, 9, 10). For the study, we used microscopic slides of 81 trees of different vitality. Sample blocks (0.5 x 0.5 x 1 cm) contained inner phloem, cambium and outer xylem taken from living trees at 1.3 m above ground during the dormant seasons of 1999 to 2003. We used observations of transverse sections using light microscopy.

Microscopic examination of cross-sections revealed that the trees could be classified into three groups on the basis of the ratios between the number of cells in the xylem and phloem growth rings (Table 1). Group 1 (43% of the trees) contained trees with up to four times more cells in the xylem ring than in phloem ones. The trees in group 2 (30% of the trees) had a ratio between xylem and phloem ring from 4.0 to 10.0, and group 3 (27% of the trees) consisted of trees with a ratio between xylem and phloem ring greater than 10.0 (Gričar et al., 2009).

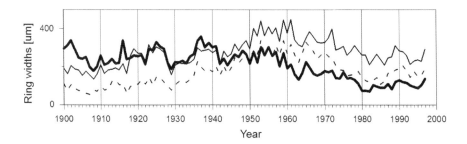

Fig. 10. Xylem ring widths of silver firs of different vitality. Category A (thin line), B (dashed line) and C (thick line) (Archives of the Chair of Wood Technology, Department of Wood Science and Technology, Biotechnical Faculty, University of Ljubljana) (Gričar, 2006).

Group	Ratio XR:PR	No. of trees (%)	No. of cell layers		Structure of PR
			XR	PR	
1	<4.0 : 1	34 (43%)	3-26	3-7	AP missing or discontinuous, EP 1-5 or > 5 cells wide, LP 1-3 cells wide or absent
2	(4.0-10.0) :1	24 (30%)	25-80	5-9	AP discontinuous or continuous, EP 2-4 cells wide, LP present
3	>10.0 :1	23 (27%)	60-144	6-12	One or two bands of AP

Table 1. Characteristics of tissues in three groups of trees with different ratios between xylem and phloem growth ring widths in terms of number of cells (XR:PR). XR – xylem growth ring, PR – phloem growth ring, AP – axial parenchyma, EP – early phloem, LP – late phloem (Gričar et al., 2009)

We confirmed that the structure and width of the phloem are closely related in silver fir (Fig. 11). Early phloem is in general 2-5 layers of cells wide and is less dependent on tree vitality whereas late phloem is subject to higher alterations in the width and type of cells. The occurrence and amount of axial parenchyma varies in accordance with the width of the phloem ring: a) it can be absent or scarce when rings are very narrow; b) present as one, more or less continuous, tangential band between early phloem and late phloem, as observed in the majority of phloem rings; or c) also forming an additional, second, discontinuous tangential band in the late phloem of very wide rings. The cambium of vital trees normally produces more xylem than phloem cells. The ratio between xylem and phloem declines with decreased vitality of trees. Only in extreme cases can the phloem ring be wider than the xylem one. The numbers of cells in phloem, xylem and dormant cambium are correlated in silver fir. Information on the width and structure of phloem rings, as well as on the relation between xylem, phloem and dormant cambium could provide additional criteria for determining tree vitality (Gričar et al., 2009).

Inspection of the current condition of the investigated trees revealed that more than half of the trees (62%) with a ratio between phloem and xylem increments lower than 4:1, with a very narrow xylem (about 20 cell layers) and phloem only 3-5 cell layers wide, died in the years following the sampling of tissues for our analyses. Our results suggest that the ratio between xylem and phloem, as well as the widths of xylem, phloem and dormant cambium, are related and indicate the health condition of a tree. They could therefore be used for assessment of the vitality of silver firs. This information could be beneficial in forest management practice, for planning the cutting of non-vital trees with poor survival prognosis and for identifying and promoting healthy and productive ones (Gričar et al., 2009).

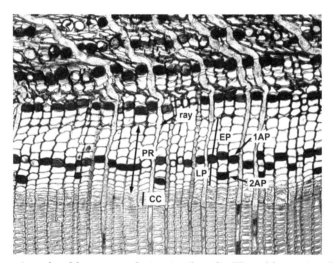

Fig. 11. Cross-section of a phloem growth ring in silver fir. PR – phloem ring, EP – early phloem sieve cells, LP – late phloem sieve cells, 1AP - first band of axial parenchyma, 2AP - second band of axial parenchyma in late phloem, CC - cambium

Since the study in silver firs gave fairly encouraging results, we tested whether similar relations can also be found in ring-porous species, such as pedunculate oak (*Quercus robur*).

11. Cambial productivity and widths of xylem and phloem increments in pedunculate oak (*Quercus robur* L.) trees of different vitality

In Slovenia, oaks (*Quercus robur* L. and *Quercus sessiliflora* Salisb.) are economically and ecologically very important wood species and represent about 7% of the entire wood stock (Zavod za gozdove Slovenije, 2010). In the case of pedunculate oak, the lowland forest area has been shrinking, due to human settlement in the past, intensive and unplanned silvicultural and agricultural exploitation of the land and conflicts of interest, so only a few lowland oak forest stands have managed to survive (Kadunc, 2010; Žibert, 2006). Similarly as in many European countries, a trend of decreasing vitality of pedunculate oak has been observed in most sites in recent decades. In 2007, pedunculate and sessile oak had the highest share of damaged and dead trees; i.e., 35.2% of the analysed tree species. The highest defoliation of pedunculate and sessile oak was observed in 2005. The condition of these species is characterised by some recuperation in 2006 and another increase in 2007 (UN-ECE, 2008).

One of the main reasons for decreasing vitality of pedunculate oak in Slovenia is ascribed to a lowering of the ground water level due to changes in climatic conditions and unsuitable artificial melioration of land for agricultural purposes (Kadunc, 2010). Namely, numerous drainage ditches were excavated in 19th century. The most obvious response of oaks to the changing environmental (hydrological) conditions is seen in the reduced wood increment, which is closely related to the structure of wood and its quality (Fig. 12).

Oak wood is considered to be very aesthetic due to its specific anatomical structure (texture) and colour. Since it also has good mechanical and durability properties, it is used for high value sawn wood products. A major factor in the utilization of wood is the degree of variation of wood properties at different scales. Variations are the result of site-to site differences in wood, population-level differences within a site and within a single tree (Gasson, 1987; Leal et al., 2007, 2008; Lei et al., 1996; Panshin & de Zeeuw, 1980; Zhang, 1997).

In addition to major economic consequences in these areas, the ecological issues associated with the decreasing vitality of pedunculate oak stands cannot be neglected. From a physiological point of view, wood tissues in trees perform several functions simultaneously, of which the two most important are to provide mechanical support and water transport. Different cell types, their morphological characteristics and their proportion in the xylem growth ring affect the survival and efficiency of the living tree. Vessel diameter, area and percentage conductive area strongly influence the amount of water that can be transported in the living tree, and so the larger the proportion of the ring occupied by conductive elements, the less tissue is available for supporting, strengthening and storage. Any changes in the proportion among different cell types therefore very likely modify the hydraulic and mechanical properties of wood (Tyree & Zimmermann, 2002).

The relationship among number of cells in phloem, xylem and dormant cambium in oak is still poorly understood. We have hypothesised that the structure and width of the phloem increments, the ratio between the phloem and xylem increments and the width of the dormant cambium would reflect the health condition of the tree. More vital trees are expected to have much wider xylem than phloem increments, whereas in declining trees, the ratio between xylem and phloem will decrease. For that purpose, we investigated the width of the phloem growth rings, late phloem, xylem growth rings, late xylem, as well as the number of cells in the dormant cambium, in 80 adult pedunculate oaks (*Quercus robur* L.) of different vitality. The health condition of the oaks was defined according to the crown condition and the width of the xylem increment.

Oak trees of various vitality were sampled at a *Pseudostellario-Carpinetum* mixed forest in Krakovo, Slovenia (45°54′N, 15 25′E, elevation 150 m). Krakovo is the largest lowland oak forest in Slovenia, which is flooded by the Krka River. It is dominated by *Quercus robur*, *Carpinus betulus* and *Alnus glutinosa* tree species. Sampled trees were dominant or co-dominant with diameter 50-60 cm, height 25-30 m and age above 80 years. In December 2009, we took micro-cores (2.4 x 2.4 x 20 mm) containing inner phloem, cambium and outer xylem taken from living trees at 1.3 m above the ground. The material extracted from the trees was immediately fixed, dehydrated in a graded series of ethanol and embedded in paraffin (Gričar, 2006). Observations and analysis were made on transverse sections using light microscopy.

The anatomical structure and widths of the phloem and xylem increments are closely related. In ring-porous oak, increasing ring width results in an increase in the percentage of late wood and late phloem, respectively (Fig. 12). In both cases, the widths of early phloem and early xylem were relatively stable as the widths of the phloem and xylem rings changed, whereas late phloem and late xylem were quite variable and increased with ring width (phloem $R2 = 0.597$; xylem $R2 = 0.955$) (Fig. 13). Other researchers have also found that the late wood portion in oak tends to increase with increased ring width, whereas the width of early wood is more or less constant (Phelps & Workman, 1994; Rao et al., 1997; Zhang, 1997). The reason is in their completely different anatomical structure and, consequently, their densities, which are much higher in late wood (ca 800 kg/m³) than in early wood (ca 560 kg/m³) (Guilley et al., 1999). Namely, the diameter of late wood vessels is much smaller and the proportion of fibers is higher. The total ring density of oak is influenced by variation in the late wood structure and by changes in the proportion of late wood to early wood.

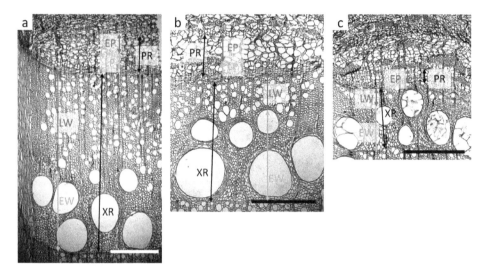

Fig. 12. Ring width and wood structure in ring-porous oaks of different vitality are closely related. XR – xylem increment, EW – early wood, LW – late wood, PR – phloem increment, EP – early phloem, LP – late phloem, Scale bars = 500 μm

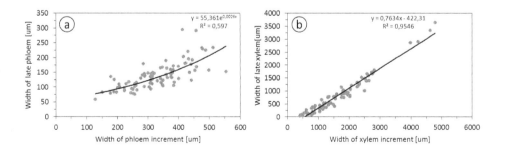

Fig. 13. Relationship between: (a) widths of late phloem and phloem increments and (b) widths of late xylem and xylem increments.

The cambium of vital trees normally produces more xylem than phloem cells. In trees with diminished vitality, xylem production is reduced and, consequently, the ratio between the xylem and phloem increment becomes progressively smaller. Only in extreme cases can the phloem increment be wider than the xylem one (Larson, 1994; Panshin & de Zeeuw, 1980). Of 86 sampled trees, the ratio between phloem and xylem in 40% of them was from 9% to 20%, in 50% of sampled oaks the ratio between phloem and xylem was from 21% to 40%, and only in 4 trees was the ratio higher than 50%. The xylem increments in this case were narrow; i.e., below 1000 μm. We found a high negative correlation between the ratio of phloem to xylem and the width of xylem increment (R2 = 0.724), whereas no such relation was found with the width of the phloem increment (Fig. 14). However, unlike in the case of silver fir, the phloem increment in pedunculate oak was smaller than the xylem one in all sampled trees.

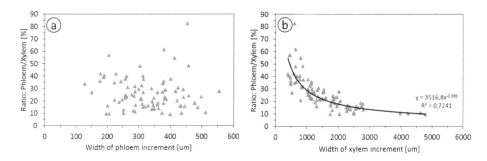

Fig. 14. Ratio between phloem and xylem increments in relation to phloem (a) and xylem (b) ring widths.

The widths of phloem and xylem and the number of cells in the dormant cambium were shown to be positively correlated. The variability in the widths of increments was higher in the xylem (400–4660 μm) than in the phloem (155-518 μm). In 25% of sampled oaks, the xylem increment was 400-1000 μm wide, in 44% 1000-2000 μm wide, in 24% 2000-3000 μm wide and in 7% 3000-4660 μm wide. In the case of phloem, 37% of oaks had an increment from 160-300 μm, 35% of them from 300-400 μm and 28% from 400-518 μm. We found a positive relationship between the width of phloem increments and the number of cells in the dormant cambium (R2 = 0.320), between the width of xylem increments and the number of cells in the dormant cambium (R2 = 0.561) and between the width of phloem and xylem increments (R2 = 0.351) (Fig. 15). The highest correlation was thus between the xylem increment and the number of cells in the dormant cambium.

We can summarize that:

- The widths of phloem and xylem increments and the number of cells in the dormant cambium are correlated in pedunculate oak.
- The widths of early phloem and early xylem are less dependent on tree vitality; whereas late phloem and late xylem are subject to higher alterations in width.
- The cambium of oak produces more xylem than phloem cells.
- The ratio between phloem and xylem increment declines with decreased vitality of trees.

The study on ring-porous pedunculate oak therefore confirms the findings obtained from coniferous silver fir trees. Information on the width and structure of xylem and phloem increments, as well as the number of cells in the dormant cambium, could indeed provide additional criteria for determining tree vitality.

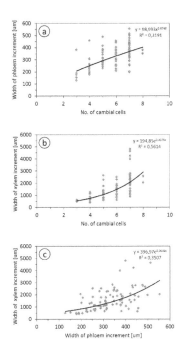

Fig. 15. Relationship between: (a) width of the phloem increment and number of cells in the dormant cambium, (b) width of the xylem increment and number of cells in the dormant cambium and (c) width of the phloem increment and width of the xylem increment in oak.

12. Conclusions

The anticipated environmental change is one of the main factors threatening the health condition of the economically most important tree species; economically essential forest stands are therefore potentially endangered. Knowledge of a species' growth characteristics and the effect that climatic variables and silvicultural management decisions have on tree growth is obviously a key issue for assessing and preserving the sustainability of forests (UN-ECE, 2008). Radial growth patterns have been shown to be valuable indicators of tree health condition. The width and structure of xylem growth rings are a source of information about past and present factors affecting development processes in an individual tree. Tree vitality also has a major effect on wood quality and properties. Information on the width and structure of xylem and phloem increments, as well as the number of cells in the dormant cambium could provide additional criteria for determining tree vitality. Indicators of tree health and vitality need to be accurate and reliable, but also cheap and easy to use. The proposed indicators comply with these desirable characteristics. Since vital forest resources are the basis of sustainable forest management and the production of quality timber, early identification of trees with an increased risk of dying could help assess and manage the health condition of forest stands in the future.

13. Acknowledgements

Thanks to Špela Jagodic from the Slovenian Forestry Institute for her invaluable help in the field and laboratory. The work was supported by the Slovenian Research Agency, programme P4-0107 and projects L7—2393 and V4—0496.

14. References

Bauch, J. (1986). Characteristics and Response of Wood in Declining Trees from Forests Affected by Pollution. *IAWA Bulletin n.s.*, Vol.7, No.4, (December 1986), pp. 269–276, ISSN 0928-1541

Beck, W. (2009). Growth Patterns of Forest Stands - The Response towards Pollutants and Climatic Impact. *iForest – Journal of Biogeosciences and Forestry*, Vol.2, (January 2009), pp. 4-6, ISSN 1971-7458

Berg, L.R. (2008). *Introductory Botany: Plants, People and the Environment*, (second edition), Thomson Brooks/Cole, ISBN-10 0534466699, ISBN-13 9780534466695, Belmont, California, USA

Bigler, C. & Bugmann, H. (2004). Assessing the Performance of Theoretical and Empirical Tree Mortality Models Using Tree-Ring Series of Norway Spruce. *Ecological Modelling*, Vol.174, No.3, (May 2004), pp. 225–239, ISSN 0304-3800

Bigler, C.; Gričar, J.; Bugmann, H. & Čufar, K. (2004). Growth Patterns as Indicators of Impending Tree Death in Silver Fir. *Forest Ecology and Management*, Vol.199, No.2-3, (October 2004), pp. 183–190, ISSN 0378-1127

Chaffey, N. (2002). Introduction, In: *Wood Formation in Trees: Cell and Molecular Biology Techniques*, N. Chaffey, (Ed.), 1-8, Taylor & Francis, ISBN-10 0415272157, ISBN-13 978-0415272155, London and New York

Coops, N.C.; Stone, C.; Culvenor, D.S. & Chisholm, L. (2004). Assessment of Crown Condition in Eucalypt Vegetation by Remotely Sensed Optical Indices. *Journal of Environmental Quality*, Vol.33, No.3, (May 2004), pp. 956-964, ISSN 0047-2425 (print version), ISSN 1537-2537 (electronic version)

Dinwoodie, J.M. (1981). *Timber, Its Nature and Behaviour*, Van Nostrand Reinhold, ISBN 0442304455, New York

Dobbertin, M. (2005). Tree Growth as Indicator of Tree Vitality and Tree Reaction to Environmental Stress: A Review. *European Journal of Forest Research*, Vol.124, No.4, (December 2005), pp. 319-333, ISSN 1612-4669 (print version), ISSN 1612-4677 (electronic version)

Ferretti, M.; König, N.; Rautio, P. & Sase, H. (2009). Quality Assurance (QA) in International Forest Monitoring Programmes: Activity, Problems and Perspectives from East Asia and Europe. *Annals of Forest Science*, Vol.66, No.4, (June 2009), pp. 403 -412, ISSN 1286-4560 (print version), ISSN 1297-966X (electronic version)

Fink, S. (1986). Microscopical Investigations on Wood Formation and Function in Diseased Trees. *IAWA Bulletin n.s.*, Vol.7, No.4, (December 1986), pp. 351–355, ISSN 0928-1541

Frampton, C.M.; Pekelharing, C.J. & Payton, I.J. (2001). A Fast Method for Monitoring Foliage Density in Single Lower-Canopy Trees. *Environmental Monitoring and Assessment*, Vol.72, No.3, (December 2001), pp. 227-234, ISSN 0167-6369 (print version), ISSN 1573-2959 (electronic version)

Franklin, J.F.; Shugart, H.H. & Harmon, M.E. (1987). Tree Death as an Ecological Process. The Causes, Consequences, and Variability of Tree Mortality. *BioScience*, Vol.37, No.8, (September 1987), pp. 550-556, ISSN 0006-3568

Fritts, H.C. (1976). *Tree Rings and Climate*, Academic Press, ISBN 0122684508 / 0-12-268450-8, London, New York and San Francisco

Gasson, P. (1987). Some Implications of Anatomical Variations in the Wood of Pedunculate Oak (*Quercus Robur* L.), Including Comparisons with Common Beech (*Fagus Sylvatica* L.). *IAWA Bulletin n.s.*, Vol.8, No.8, (June 1987), pp. 149–166, ISSN 0928-1541

Ghosh, S. & Innes, J.L. (1995). Combining Field and Control Team Assessments to Obtain Error-Estimates for Surveys of Crown Condition. *Scandinavian Journal of Forest Research*, Vol.6, No.1-4, (January 1995), pp. 264-270, ISSN 0282-7581 (print version), ISSN 1651-1891 (electronic version)

Gorišek, Ž. (2009). *Les : zgradba in lastnosti : njegova variabilnost in heterogenost*, Reviewed University and Academic Textbook, Department of Wood Science and Technology, Biotechnical Faculty, University of Ljubljana, ISBN 978-961-6144-28-5, Ljubljana, Slovenia, (in Slovenian language)

Gričar, J. (2006). *Effect of Temperature and Precipitation on Xylogenesis in Silver Fir (Abies Alba) and Norway Spruce (Picea Abeis)*, Doctoral Dissertation, Department of Wood Science and Technology, Biotechnical Faculty, University of Ljubljana, Ljubljana, Slovenia, (in Slovenian language)

Gričar, J. & Čufar, K. (2008). Seasonal Dynamics of Phloem and Xylem Formation in Silver Fir and Norway Spruce as Affected by Drought. *Russian Journal of Plant Physiology*, Vol.55, No.4, (July 2008), pp. 538-543, ISSN 1021-4437 (print version), ISSN 1608-3407 (electronic version)

Gričar, J. (2009). Significance of Intra-Annual Studies of Radial Growth in Trees, In: *TRACE - Tree Rings in Archaeology, Climatology and Ecology, Volume 7: Proceedings of the Dendrosymposium 2008, April 27th - 30th, 2008 in Zakopane, Poland*, R.J. Kaczka et al., (Eds.), 18-25, GFZ German Research Centre for Geoscience, ISSN 1610-0956, Potsdam, Germany

Gričar, J.; Krže, L. & Čufar, K. (2009). Relationship among Number of Cells in Xylem, Phloem and Dormant Cambium in Silver Fir (*Abies Alba* Mill.) Trees of Different Vitality. *IAWA Journal*, Vol.30, No.2, (June 2009), pp. 121-133, ISSN 0928-1541

Groover, A.T. (2005). What Genes Make a Tree a Tree? *Trends in Plant Science*, Vol.10, No.5, (May 2005), pp. 210-214, ISSN 1360-1385

Guilley, É.; Hervé, J.-C.; Huber, F. & Nepveu, G. (1999). Modelling Variability of Within-Ring Density Components in *Quercus Petraea* Liebl. with Mixed-Effect Models and Simulating the Influence of Contrasting Silvicultures on Wood Density. *Annals of Forest Science*, Vol.56, No.6, (September 1999), pp. 449-458, ISSN 1286-4560 (print version), ISSN 1297-966X (electronic version)

Innes, J.L. (1993). Methods to Estimate Forest Health. *Silva Fennica*, Vol.27, No.2, (June 1993), pp. 145-157, ISSN 0037-5330

Innes, J.L. (1993). Some Factors Affecting the Crown Condition Density of Trees in Great Britain Based in Recent Annual Surveys of Forest Condition, In: *Forest Decline in the Atlantic and Pacific Region*, R.E. Huettl & D. Mueller-Dombois, (Eds.), 40-53, Springer-Verlag, ISBN 3540546405, Berlin-Heidelberg

Kadunc, A. (2010). Quality, Value Characteristics and Productivity of Pedunculate and Sessile Oak stands in Slovenia. *Gozdarski vestnik*, Vol.68, No.4, (May 2010), pp. 217-226, ISSN 0017-2723, (in Slovenian language)

Kolb, T.E.; Wagner, M.R. & Covington, W.W. (1994). Concepts of Forest Health: Utilitarian and Ecosystem Perspectives. *Journal of Forestry*, Vol.92, No.7, (July 1994), pp. 10-15, ISSN 0022-1201

Kozlowsky, T.T. & Pallardy, S.G. (1997). *Growth Control in Woody Plants*, Academic Press, Inc. ISBN 0-12-424210-3, San Diego, California

Lachaud, S.; Catesson, A.M. & Bonnemain, J.L. (1999). Structure and Functions of the Vascular Cambium. *Comptes Rendus de l'Académie des Sciences. Sciences de la Vie (Life Sciences)*, Vol.322, No.8, (August 1999), pp. 633-650, ISSN 0764-4469

Larcher, W. (2003). *Physiological Plant Ecology: Ecophysiology and Stress Physiology of Functional Groups*, (fourth edition), Springer–Verlag, ISBN 3-540-43516-6, Berlin-Heidelberg-New York

Larson, P.R. (1994). *The Vascular Cambium: Development and Structure*, Springer–Verlag, ISBN 3540571655, Berlin-Heidelberg-New York

Leal, S.; Sousa, V.B. & Pereira, H. (2007). Radial Variation of Vessel Size and Distribution in Cork Oak Wood (*Quercus Suber* L.). *Wood Science and Technology*, Vol.41, No.4, (April, 2007), pp. 339-350, ISSN (print version), ISSN 1432-5225 (electronic version)

Leal, S.; Nunes, E. & Pereira, H. (2008). Cork Oak (*Quercus Suber* L.) Wood Growth and Vessel Characteristics Variations in Relation to Climate and Cork Harvesting. *European Journal of Forest Research*, Vol.127, No.1, (January 2008), pp. 33-41, ISSN 1612-4669 (print version), ISSN 1612-4677 (electronic version)

Lei, H.; Milota, M. R. & Gartner, B.L. (1996). Between- and Within-Tree Variation in the Anatomy and Specific Gravity of Wood in Oregon White Oak (*Quercus Garryana* Dougl.). *IAWA Journal*, Vol.17, No.4, (December 2003), pp. 445-461, ISSN 0928-1541

Legg, C.J. & Nagy, L. (2006). Why Most Conservation Monitoring Is, But Need Not Be, a Waste of Time. *Journal of Environmental Management*, Vol.78, No.2, (January 2006), pp. 194-199, ISSN 0301-4797

Martín-García, J.; Diez, J.J. & Jactel, H. (2009). Towards Standardised Crown Condition Assessment in Poplar Plantations. *Annals of Forest Science*, Vol.66, No.3, (April-May 2009), pp. 308, ISSN 1286-4560 (print version), ISSN 1297-966X (electronic version)

Mauseth, J.D. (1988). *Plant Anatomy*, Benjamin/Cummings Publishing Company, ISBN 0805345701 Menlo Park, California

Mizoue, N. (2002). CROCO: Semi-Automatic Image Analysis System for Crown Condition Assessment in Forest Health Monitoring. *Journal of Forest Planning*, Vol.8, No.1, (April 2002), pp. 17-24, ISSN 1341562X

Mizoue, N. & Dobbertin, M. (2003). Detecting Differences in Crown Transparency Assessments between Countries Using the Image Analysis System CROCO. *Environmental Monitoring and Assessment*, Vol.89, No.2, (December 2003), pp. 179-195, ISSN 0167-6369 (print version), ISSN 1573-2959 (electronic version)

Panshin, A.J. & de Zeeuw, C. (1980). *Textbook of Wood Technology*, (fourth edition), McGraw-Hill, ISBN 0070484414, New York

Phelps, J.E. & Workman, E.C. (1994). Vessel Area Studies in White Oak (*Quercus Alba* L.). Wood and Fiber Science, Vol26, No.3, (July 1994), pp. 315-322, ISSN 07356161

Plomion, C.; Leprovost, G. & Stokes, A. (2001). Wood Formation in Trees. *Plant Physiology*, Vol.127, No.4, (December 2001), pp. 1513–1523, ISSN 0032-0889 (print version), ISSN 1532-2548 (electronic version)

Rao, R.V.; Aebisher, D.P. & Denne, M.P. (1997). Latewood Density in Relation to Wood Fibre Diameter, Wall Thickness, and Fibre and Vessel Percentages in *Quercus Robur* L. *IAWA Journal*, Vol.18, No.2, (June 1997), pp. 127-138, ISSN 0928-1541

Savidge, R.A. (2001). Intristic Regulation of Cambial Growth. *Journal of Plant Growth Regulation*, Vol.20, No.1, (March 2001), pp. 52-77, ISSN 0721-7595 (print version), ISSN 1435-8107 (electronic version)

Schmitt, U.; Grünwald, C.; Gričar, J.; Koch, G. & Čufar, K. (2003). Wall Structure of Terminal Latewood Tracheids of Healthy and Declining Silver Fir Trees in the Dinaric Region, Slovenia. *IAWA Journal*, Vol.24, No.1, (March 2003), pp. 41-51, ISSN 0928-1541

Schweingruber, F.H. (1986). Abrupt Growth Changes in Conifers. *IAWA Bulletin n.s.*, Vol.7, No.4, (December 1986), pp. 277–283, ISSN 0928-1541

Shigo, A.L. (1986). *A New Tree Biology*, Shigo & Trees Associates, ISBN 0943563127, Durham, New Hampshire

Solberg, S. & Strand, L. (1999). Crown Density Assessments, Control Surveys and Reproducibility. *Environmental Monitoring and Assessment*, Vol.56, No.1, (May 1999), pp. 75-86, ISSN 0167-6369 (print version), ISSN 1573-2959 (electronic version)

Stone, C.; Wardlaw, R.F.; Carnegie, A., Wyllie, R. & De Little, D. (2003). Harmonization of Methods for the Assessment and Reporting of Forest Health in Australia – A Starting Point. *Australian Forestry*, Vol.66, No.4, (December 2009), pp. 233-246, ISSN, 0004-9158

Taiz, L. & Zeiger, E. (2002). *Plant Physiology*, (third edition), Sinauer Associates Inc. Publishers, ISBN 0-87893-823-0, Sunderland, Massachusetts

Torelli, N.; Čufar, K. & Robič, D. (1986). Some Wood Anatomical, Physiological and Silvicultural Aspects of Silver Fir Dieback in Slovenia. *IAWA Bulletin n.s.*, Vol.7, No.4, (December 1986), pp. 343–350, ISSN 0928-1541

Torelli, N.; Shortle, W.C.; Čufar, K.; Ferlin, F. & Smith, K.T. (1999). Detecting Changes in Tree Health and Productivity of Silver Fir in Slovenia. *European Journal of Forest Pathology*, Vol.29, No.3, (June 1999), pp. 187–197. ISSN 0300-1237 (print version), ISSN 1573-846 (electronic version)

Tyree, M.T. & Zimmermann, M.H. (2002). *Xylem Structure and the Ascent of Sap*, Springer–Verlag, ISBN 3-540-43354-6, Berlin-Heidelberg-New York

UN-ECE (2008). The Condition of Forests in Europe: 2006 Executive Report. Institute for World Forestry, ISSN 1020-587X, Hamburg, Germany

Vos, P.; Meelis, E. & Ter Keurs, W.J. (2000). A Framework for the Design of Ecological Monitoring Programs as a Tool for Environmental and Nature Management. *Environmental Monitoring and Assessment*, Vol.61, No.3, (April 2000), pp. 317–344, ISSN 0167-6369 (print version), ISSN 1573-2959 (electronic version)

Zavod za gozdove Slovenije. (2010). Poročilo zavoda za gozdove Slovenije o gozdovih za leto 2009. Ljubljana, (in Slovenian language), 14.07.2011; Available from http://www.zgs.gov.si/fileadmin/zgs/main/img/PDF/LETNA_POROCILA/Porgozd10_Solc1.pdf

Zhang, S.Y. (1997). Variations and Correlations of Various Ring Width and Ring Density Features in European Oak: Implications in Dendroclimatology. *Wood Science and Technology*, Vol.31, No.1, (February 1997), pp. 63–72, ISSN (print version), ISSN 1432-5225 (electronic version)

Waring, R. H.; Thies, W. G. & Muscato, D. (1980). Stem Growth per Unit of Leaf Area: A Measure of Tree Vigour. *Forest Science*, Vol.26, No.1, (March 1980), pp. 112-117, ISSN 0015-749X

Waring, R. H. (1987). Characteristics of Trees Predisposed to Die. *BioScience*, Vol.37, No.8, (September 1987), pp. 569-574, ISSN 0006-3568

Wimmer, R. (2002). Wood Anatomical Features in Tree-Rings as Indicators of Environmental Change. *Dendrochronologia*, Vol.20, No.1-2, (January 2002), pp. 21–36, ISSN 1125-7865

Wodzicki, T.J. (2001). Natural Factors Affecting Wood Structure. *Wood Science and Technology*, Vol.35, No.1-2, (April 2001), pp. 5-26, ISSN (print version), ISSN 1432-5225 (electronic version)

Žibert, F. (2006). *Stand Structure in Virgin Forest Reserve Krakovo and Managed Forest*, Graduation Thesis - Higher Professional Studies, Department of Forestry and Renewable Forest Resources, Biotechnical Faculty, University of Ljubljana, Ljubljana, Slovenia, (in Slovenian language)

7

A Common-Pool Resource Approach to Forest Health: The Case of the Southern Pine Beetle[*]

John Schelhas[1] and Joseph Molnar[2]
[1]Southern Research Station, USDA Forest Service
G.W. Carver Agricultural Experiment Station, Tuskegee University, Tuskegee
[2]Department of Agricultural Economics and Rural Sociology
Alabama Agricultural Experiment Station, Auburn University
USA

1. Introduction

The southern pine beetle, *Dendroctonus frontalis*, is a major threat to pine forest health in the South, and is expected to play an increasingly important role in the future of the South's pine forests (Ward and Mistretta 2002). Once a forest stand is infected with southern pine beetle (SPB), elimination and isolation of the infested and immediately surrounding trees is required to control the outbreak. If insect-infested trees are not swiftly removed, infestations can spread to healthy forests. The most effective approach to managing SPB is through preventive measures that maintain forests in vigorous, healthy conditions, including thinning and prescribed burning. At a landscape level, preventive measures reduce the overall incidence of SPB and thereby the spillover of SPB to adjacent landholdings. Yet many forest landowners do not undertake the management actions that can limit SPB outbreaks. The tragedy of the commons in forest health takes place when individual private owners do not acknowledge their communal responsibilities thus risking catastrophic losses due to poor management and/or absentee tenure.

The South's forests are largely in private ownership (89% of the South's timberland, with nonindustrial private forest (NIPF) land ownerships representing about 95% of the private forest landowners and 63% of the private forest land region (Birch 1996, Wicker 2002). Population growth and suburban and exurban expansion in the South have divided many forest landholdings into increasingly smaller-sized parcels. Surveys of forest landowners in the South find that 90% of the NIPF owners hold less than 100 acres, and that owners are diverse in occupation, income, residence, forest land ownership objectives, use of professional forest management assistance, and forest management strategies (Birch 1996, 1997; Bliss and Martin 1989).

The diversity of ownership objectives and management styles on NIPF lands results in widely different awareness and responses to forest pest problems (Ward and Mistretta 2002). Pine beetle outbreaks are cyclic, sporadic, and potentially highly devastating (Meeker

---

[*] Paper presented to the 67th Annual Meeting of the Southern Sociological Society, 14-17 April, Atlanta. Research supported the U.S. Forest Service and the Alabama Agricultural Experiment Station.

et al. 1995). Extensive outbreaks not only inflict setbacks on individual owners who suffer losses from forced sale of high-value saw timber for low-value pulp, but also collective damages on all forest owners.

The maintenance of healthy pine forests and the various benefits associated with them in the South depends on effective management and control of the Southern Pine Beetle. To a significant extent, SPB management is a social problem because the most practical way to control SPB requires collective action by individual landowners across the pine forest landscapes in the South. Most social research on programs for forest landowners in the U.S. has tended to view them as individuals, and be oriented toward transferring new knowledge, technical assistance, financial assistance and even cultural content to autonomous forest landowners (Best and Wayburn 2001; Schelhas et al. 2004). Accordingly, we have oriented much of our analysis on forest landowners and SPB to understanding why individual landowners do or do not engage in practices known to be effective in the prevention of SPB (Molnar et al. 2003).

However, we also recognize that, from a social science viewpoint, the characteristics of the SPB issue–the need for action at the landscape level, when landscapes are in multiple ownerships--is a problem of the commons (Ostrom 1990). Natural resource management in the commons has been subject to a great deal of study over the past few decades, although little or none of this research has addressed questions of forest health. However we believe that the general principles of the management of common-pool resources can provide some important insights for SPB management. In this paper we explore the usefulness of examining the management of SPB from the perspective of common-pool resource management. As Hardin (1968) notes, an implicit and almost universal assumption of discussions of resource management problems is that a technical solution must exist and the task is to find it. A technical solution may be defined as one that requires a change only in the techniques of the material sciences, demanding little or nothing in the way of change in human values or ideas of morality.

2. A brief review of theory of common-pool resources and forests

Three types of resources can be identified based on different combinations of two characteristics: (1) *subtractability or rivalness*, or the degree to which use by one person diminishes the potential for use by another, and (2) *excludabilty*, the cost of excluding potential beneficiaries from the resource (McKean 2003). **Private** resources are subtractable in consumption and others can be excluded relatively easily. **Public** resources are available to all (exclusion is not possible or is extremely costly) but not subtractable. Examples include public radio stations, scientific knowledge, and world peace. Individuals may enjoy the benefits of these without contributing to their production (free ride), but if everyone does this a less than ideal amount of the good will be provided (Dietz 2001, Ostrom and Walker 1997). **Common-pool** resources are subtractable but exclusion is difficult (Dolsak and Ostrom 2003).

Although it has been common in the past to discuss common property resources, recent work has emphasized the importance of distinguishing types of resources (based on their inherent attributes, from types of ownership (Dietz et al. 2001). Property may be held in four ways: (1) *private*, in which individuals or corporations have the rights to exclude others from using a resource and to regulate a resource; (2) *public or state*, in which the government has rights to a resource, and makes decisions about access as well as the nature and level of exploitation; (3) *common property*, in which the resource is held by an identifiable group of interdependent users with the rights to exclude others, and (4) *open access*, in which there are no well-defined property rights, the resource is unregulated, and it is free and open to everyone (Feeny et al. 1990). Research on the commons suggests that the fit between property type and resource type has an important bearing on effective resource management (Dietz et al. 2001, Stern et al. 2002).

Geores (2003) points out that forests are complex, large scale resources that can be defined and assigned property rights in various ways: (1) Forest are appreciated as renewable natural resources, valued for the use of their products and for their roles in maintaining watersheds, soil fertility, and air quality, as well as for their importance as cultural resources, both religious and aesthetic. (2) Forests are resources that contain resources, being made up of biosystems of varying complexity and used for many different social and economic functions as a part of complex social systems. (3) Forests resources are dynamic and defined on multiple scales. Forest and forest resource definitions differ in scale, but are not necessarily mutually exclusive.

Southern forests illustrate this in the way that the wider public values them for wildlife, watershed, biodiversity, and climatic benefits (each requiring management at different scales). In contrast, trees and forests are used and valued by individual landowners for timber. Even when considering only a single resource, such as timber production or wildlife by individual owners, owners of individual parcels may want to encourage or guarantee that owners of adjacent parcels have compatible and complementary interests in their parcels. Neighbors want their neighbors to maintain wildlife habitat and keep vegetative cover intact. They also want adjacent land owners to allow wildlife transit and to refrain from introducing or encouraging certain problem species (McKean 2000).

Gibson and Becker (2000), recognizing that forests generally constitute multiple resources, note that strong individual property rights alone do no guarantee a forest's health since individuals can have short term incentives to convert or degrade forests that conflict with long term forest sustainability. Because they are common-pool and public resources, many forest resources cannot be effectively managed on the scale at which they are owned or in the decision-making time frames of some private owners. As a result, individual forest owners have an interest in what happens on lands adjacent to theirs. Southern pine beetle is a classic example of the stake neighbors have in the way their neighbors attend to forest health.

One of the problems facing common-pool resources is the appropriation problem. If resource units have high value and institutional constraints do not restrict use, individuals face a strong temptation to overexploit and thereby degrade the resource. For example if a forest is open to access by all with no social institutions to limit use, is it likely that timber would be removed at such a rate that the forest would degrade and future timber harvests would be reduced (Hardin 1968). Extensive study of the appropriation problem by social scientists has found that the tragedy of the commons is not inevitable; resource users can organize to implement social mechanisms to restrict use to sustainable levels (Richard and Stein 2003). Other problems of common-pool resources, such as provision and maintenance problems, have received less study but are still important (Ostrom 1999).

Forest health is essentially a provision and maintenance problem. In many ways, it is a public good, in that people can free ride on other people's efforts to enhance forest health at a landscape or regional level. But McKean (2003) notes that public goods that are subject to crowding, wear, and depletion are not pure public goods, and have many characteristics of common-pool resources. Furthermore, Ostrom (1999) notes that in the case of negative public goods (e.g. forest pests), individual owners or appropriators tend not to be motivated to pay for or take the collective actions that are required to reduce the negative public good, resulting in a negative provision of that good (e.g. poor forest health). Provision problems in common-pool resources are very similar to pure public good problems (Ostrom et al. 1994).

Having shown that forests and forest health have important attributes of common-pool resources, the next question is what common-pool resource theory and scholarship can contribute to the health and management of Southern pine forests. Ostrom and Walker (1997) examined many cases of successful common-pool resource management. They identified design principles for development of institutions that increase the efficiency of management of common-pool resources, institutions that are often developed in combination by the resource users and the state.

3. Key understandings from research on individual NIPF owners

A legacy of medieval times, Carlsson (1996) explains why Swedish common forests have survived as vital and competitive actors in the timber market. These lands are held in common under shareholder arrangements managed by the government. He offers three main explanations: the commoners' conscious attempts to reduce transaction costs, their general inventiveness in adjusting to changed circumstances, and their acclimatization to present economic conditions. Although he does not specifically address forest health issues, the notion that a commons institutions offers multiple advantages to a dispersed, nonresidential, and nontechnical population of forest owners suggests a need for new institutions and mechanisms to bind and benefit nonindustrial private forest land owners (NIPF).

Most NIPF landowners are aware of SPB, many are interested in preventing the pest, and some express a desire to accomplish control measures (Molnar et al. 2003). Those actually taking action to prevent and manage infestations are few, however.

Molnar et al. (2003) found important differences by size of forest landholding. Larger landholders are more likely to have taken steps to control infestations, but there were markedly lower levels of awareness, surveillance, and prevention activities among small holders. Larger landowners had high surveillance efforts and took more action to respond to SPB damage when it happened on their land. Larger landowners were also strongly influenced by timber prices in their efforts to control SPB.

Smallholders lacked knowledge about what to do about SPB, lacking familiarity with public agency programs and utilization of financial assistance. They used fewer information sources, and expressed less desire for information about forest management (Molnar et al. 2003).

Some values that landowners–large and small–have for their forest land may provide less than compelling motivations for SPB management. Those interested in recreation and outdoor enjoyment and indicating preservation as a primary reason for forest ownership were less aware and interested in SPB management (Molnar et al. 2003). The control of SPB and the protection of forest health, involves more than the vigilance of the individual forest owner, however.

Carlsson (1996: 12) concludes that the Swedish forest commons have survived as prosperous timber producers and providers of public goods, not only because of their conscious reduction of transaction costs but also because this reduction has been made possible by a general fragmentation of the centralized State, playing its multiple roles. This fragmentation has provided a local 'opportunity structure' that the commons have utilized. This has been possible because the commons, their forest managers, boards and assemblies of shareholders still possess sufficient local, current knowledge to be able to adjust the commons to industrialized society. The main lesson to be learned from the Swedish common forests might be their successful integration, rather than their separation, from the logic of the negotiated economy and industrialized society. Designers of institutional mechanisms to articulate and organize the collective aspects of forest health might learn much from the Swedish experience.

4. Calculating the benefit from change in rules of forest management

Ostrom (1999:4) emphasizes that the "social behavior of adopting new practices in natural resources management as a rational decision process. Each user has to compare the net benefits continuing to use the old rules of harvesting from a resource to the benefits he or she expects to achieve with a new set of rules. Each user must ask whether his or her incentive to change is positive or negative.

If the incentive to change is positive for some users, they then need to estimate three types of costs: the up-front cost of time and effort devising and agreeing upon new rules; the short-term costs of adopting new strategies, and the long-term cost of monitoring and maintaining a self governed system over time (given the norms of community where they live). If the sum of these expected costs for each user exceeds the incentive to change, no user will invest the time and resources needed to create new institutions. And if this applies to all the users, no change will occur (Ostrom 1999:4).

In field settings, not everyone expects the same cost and benefits from a proposed change. Consequently, the collective choice rules used to change the day-to-day operational rules related to management activities affect whether an institutional change favored by some and opposed by others will occur (Ostrom 1999:4).

These comparisons can be difficult to make in practice since considerable uncertainty always exists concerning the strategies that participants will follow once rules are changed (Ostrom 1999:4). But even though this is a difficult task, it is one undertaken frequently by users after discussing the effects of a change in rules. Rules about monitoring forest lands for SPB infestation may be one example of an institutional change.

Prevention efforts require vigilant surveillance for infestations and adherence to planting and management recommendations that discourage SPB outbreaks. Once outbreaks occur, control requires prompt treatment, and a comprehensive response by all forest owners to stop the spread of SPB to neighboring lands (Egan and Jones 1993, Ervin et al. 2001). Yet many NIPF owners have weak and uneven ties to their properties, and many do not share the sense of urgency that professional foresters often have about SPB prevention and control (Williston et al. 1998).

5. Forest health as a common property resource

Land (and forest) tenure is now widely understood as bundle of rights, all or some of which may be privately owned. Under communal systems, no individual resource rights are privately owned. Under private property systems, the deed holder seemingly owns all rights.

It is increasingly clear that some rights in the bundle can never be exclusively held by individuals, and are in fact dependent on communal cooperation and respect. Forest health may be one such communally owned and managed resource that is held by all forest owners but no one singly. This common pool, open access resource, abused by one, can cause all to suffer. An ephemeral and situational commodity, forest health is often taken for granted when insects, fire, or other threats are not imminent.

The owners of the forest health right or resource are connected in concentric levels of proximity. That is, near neighbors are more frequently and intensively affected by mutual actions and responsibilities. Distant parties are less frequently benefited or harmed by an individual landowner's vigilance and response to forest health problems. Institutions such as forest fire districts sometimes connect land owners in defense of fire threats, but fire threats are not commonly limited to pest prevention.

These indirect and fleeting communal connections among NIPF owners are at the core of the problems facing public agencies charged with promoting forest health. For the most part, locally resident forest land owners often have little basis for interpersonal association. Even among landowners who reside in the same county as their forest land, the increasing separation of residence from ownership diminishes the prospect for face-to-face interaction with neighboring forest land owners.

McKean and Ostrom (1995) find it noteworthy that the definition of private property rights has to do with the rights, not the nature of the entity that holds them. The privateness of private property rights does not require that individual persons hold them; they may also be vested in groups of individuals. Unfortunately, the rights to forest health are not alienable or separable; such rights are evanescent or intangible. Yet when unevenly exercised, forest fires or large-scale timber losses from insect damage are the result.

Scholars who have designed taxonomies to point out the difference between open access arrangements and common property have sometimes distinguished four very general "types" of property: public, private, common, and open access. McKean and Ostrom (1995) object to this classification because it creates the erroneous impression that common property is not private property and thus does not share in the desirable attributes of private property, although forest health property rights are indeed commonly held. They feel that common property is in fact shared private property and should be considered alongside business partnerships, joint-stock corporations and cooperatives. Yet, the shared resource of forest health is often not widely recognized as a common property resource

Oakerson (1986) has suggested a model to analyze and explain the main factors involved in the management of common property resources. In its simplest form, the Oakerson model is based on understanding the relationships between the physical characteristics of the resource, the decision making rules of the group or users involved, the patterns of interactions resulting from the appropriation and use of the resource, and the outcomes of this process. Blaikie and Brookfield (1987) have modified the Oakerson model to explain the dynamic interactions and adaptive changes when a resource is managed under a communal (or collective) regime.

Mutual regulation through the institutional equivalent of a common property regime is more desirable as resource use intensifies and approaches the productive limits of a resource system (McKean and Ostrom 1995). Further, since it is people who use resources, forest health common property becomes more desirable - not necessarily more workable but more valuable and thus more worth trying - as population density increases on a given resource base. Thus the challenge to resource agencies endeavoring to create a common property resource in forest health must find a way to communicate with NIPF owners in such a way so they become aware of the common property resource they share and have a sense of ownership in the commons.

Natural resources stakeholders have different interests, and investigation of these through discussion can help to identify how people view their current and potential roles in forest management (Higman et al.1999: 170). The challenge to resource managers is to communicate the common property resource aspects of forest health. Higman et al. (1999:170) claim that finding out how people see their own roles in forest management is an essential step toward agreeing about the objectives of forest management. One way of doing this is to focus discussion on stakeholders' rights, responsibilities and results with respect to surveillance and timely response to SPB outbreaks.

As a result from their different rights, responsibilities and returns, stakeholders also have different sorts of relationships with each other. Some may not be aware of each other, or may ignore each other; others may be in varying states of disagreement or cooperation in different issues related to forest management. Yet all share some level of common interest in forest health.

6. Characterizing a robust common property system for forest health

A robust system of social organization for NIPF owners that would promote and protect the common property aspects of forest health has yet to be devised. McKean (1992, 1996, 2000) has written on the nature of common property systems that would lead to ecological benefits for the natural world. She identifies a number of design criteria that may make common property systems robust (McKean 2000a), focusing on internal and external features of the resource management system.

Internal Features pertain to relationships among co-owners, that is, among NIPF owners. Each of McKean's design features is discussed in terms of a common property management system for forest health.

1. Co-owners of resource rights must be a self-conscious and self-governing group.

This feature is hard to envision occurring beyond a watershed or county scale. As previously discussed, nonresident, nontechnical, and dispersed landowners have no mechanism for communication or collaboration. Thus efforts to promote the common pool resource aspects of forest health must develop new mechanisms for linking heretofore-unconnected NIPF owners.

2. The group needs a mechanism for resolving internal conflict.

Current mechanisms generate little direct conflict because NIPF owners have little occasion to interact with one another. Animosity toward noncompliant landowners may be manifested under specific circumstances, but the forest health consequences of NIPF owner indifference or neglect are typically absorbed or ignored by neighboring landowners.

3. The rules need to provide for monitoring of behavior and enforcement of sanctions.

Some states have laws and regulations that sanction noncompliant NIPF for neglecting SPB infestations, yet it is not clear how often these measures are put into play nor how effective they are in influencing behavior.

4. The rules need to include arrangements to prevent abuse by guards.

It is not clear who the "guards" might be for forest health. At present, public forest managers monitor aerial photos and accumulate reports of infestations to provide assessments of SPB problems. Under a common property regime, NIPF owners themselves might play a greater role in surveillance, requiring access to private lands and other measures that might otherwise compromise individual property rights. If such access were used for private gain – e.g., off-roading, hunting, fishing, or trapping -- cooperation and the common property institution would be undermined.

5.   **The rules need to be easily enforceable and ecologically conservative.**

Rules for managing forest health as common property would require a great deal of public education and would have to be nested in the current web of property law and public agency regulation. Monitoring and infestation response requirements would have to achieve a level of technical and sociopolitical consensus about the techniques of SPB control.   Motivating NIPF owners to participate in such discussions would a challenge to resource management agencies not only in terms of the sheer number of actors that would have to be contacted, but also in terms of the communication and participation efforts that would be needed to enlist and sustain NIPF owner involvement and commitment.

6.   **The allocation of benefits from the commons needs to be roughly proportional to the effort (time, money) invested in the commons.**

Under the Swedish system discussed earlier, common members are shareholders in corporate institutions that protect and manage production from forest lands (Carlsson 1996). A U.S. system that endeavored to enlist NIPF in monitoring and managing forest health on a per acre basis might not produce sufficient incentives for small holders. Devising institutional incentives that motivate participation and commitment from large and small holders would have to balance the costs of participation with the infrequently tangible, usually delayed, and often diffuse benefits of forest health.

External Features encompass relationships between the body of co-owners and the outside world. Four considerations relate to the issue of forest health.

7.   **The co-owning community of resource users is much better off if it has independent jurisdiction or autonomy**

Soil and water conservation districts are examples of communities of resource users that have some independent jurisdiction. Such entities are, however, creatures of state and federal laws that enable them. It is clear that not all landowners participate, nor do all that participate benefit equally from these programs – particularly in terms of size of holding and ethnicity of the land owner (Schelhas 2003).

8.   **The boundaries of common property regimes need to be set at an appropriate ecological** scale and need to match ecosystem boundaries.

It is not clear what the appropriate ecological scale is for forest health. Other efforts are underway to organize land owners on the scale of the watershed, thus is seem prudent to seek coincident boundaries between soil, water, and forest resource units of social organization. McKean (2000:10) points out that it is silly to introduce common property institutions where parceled individual property would make more sense, and it is vital to use common property where parcelization to individuals is not a good idea. Forest health is not a resource that is easily parcellized.

9.   **It is important to select the right group to vest common property rights in order to get capacity to affect the problem.**

The unit of organization must be close enough to the problem to aggregate individual decisions and realize consequences for the resource to be managed. A common property institution should combine NIPF forest landowners in a way that connects their efforts to the cause of forest health and achieves demonstrable consequences for the resource as well as the NIPF owners.

10. On large resource systems, it is important to nest new layers of governance (federalism)

Social organization designed to coalesce NIPF owners to achieve forest health must be aligned with the other emerging forms of association that endeavor to promote and protect resources. Forest resource management must complement water and soil management efforts; there must be some level of mutual reinforcement and synergy to achieve effective environmental management. The environment is interconnected; so must the efforts to make it sustainable.

7. Social capital, social organization, and common property

Each U.S. county has some level of social capital – fire districts, irrigation districts, soil conservation districts, forest associations, extension councils, etc.--that can be drawn on to construct the common property institution in forest health. Institutional changes that expand fire protection vigilance to forest health surveillance including SPB monitoring can build on existing social arrangements to protect forest health. Flora (2000: 87) notes the importance of building human and social capital for communities that are engage in natural resources management. Social capital involves mutual trust where people know they can count on someone, which fosters reciprocity. Mutual trust is established when different institutions and individuals can both give and receive. Mutual trust and reciprocity tend to occur when people work together.

Flora (2000: 87) mentions that one way of building trust is to start with small projects that have immediate visible results that everyone can measure and contribute to. Face-to-face groups are the building blocks of social capital. The measurement of increased social capital is done by looking at the strengthened relationships and communication among unlikely segments within or outside the community and the increased availability of information and knowledge. McDonald and McLain (2003) describe the successful integration of community well-being and forest health in the Pacific Northwest. They found that a central vehicle for change was the creation of a quasi-public organization (Conservation and Development Council) that had as its first objective to improve economic and social well-being. Specifically, the Council promoted forest health and community well-being through habitat restoration programs that employed people in the area. The Council used special forest products programs to encourage businesses to pool resources for equipment and marketing, and give employees training in forest products harvest and marketing. The Council also sponsored a wood products production and marketing activities programs to help public and private owners produce and market wood products.

Council activities also played an important role in creating new alliances and changing relationships among local and non-local organizations. It increased the capacity of local groups to obtain funds and gain access to technical expertise from outside organizations. In short, it provided an institutional substrate for managing the forest health commons.

Ostrom (1999:2) defined a self-governed forest resource as one where actors, who are major users of the forest, are involved over time in making and adapting rules within collective-choice arenas regarding the inclusion or exclusion of participants, appropriation strategies, obligation of participants, monitoring or sanctioning, and conflict resolution. In most modern political economies it is rare to find any resource system that are governed entirely by participants without rules made by local, regional and national authorities also affecting key decisions. Thus in a self-governed system, participants make many, but not all, rules that affect the sustainability of the resource system and its use.

Both the natural physical boundaries of a forest as well as the legal boundaries for a particular community's forest must be clearly identified and defined (McKean and Ostrom 1995). The lack of definition and assignment of forest health property rights quite clearly represents a barrier to forestry management, on the one hand limiting the realization of prevention and control benefits and, on the other, encouraging "free rider" behavior and giving rise to the so-called tragedy of the commons — outbreaks that spread to neighboring properties and create otherwise avoidable catastrophic timber losses.

Inflexible rules are brittle, and thus fragile, and can jeopardize an otherwise well-organized common property regime (McKean and Ostrom 1995). In particular, the science behind SPB had not fully defined the rise and fall of SPB populations. Consequently, in some years natural forces driving surges in SPB infestation may overcome high levels of surveillance and response to outbreaks. The setbacks and frustrations occurring to NIPF owners stress the institutions that normally prevent and control SPB outbreaks.

Institutions for managing very large systems need to be layered, with considerable authority devolved to small components. Many different communities, some of which are in frequent contact with each other and some of which are not, may use a large forest. The need to manage a large forest as a unit would seem to contradict the need to give each of that forest's user communities some degree of independence. Nesting different user groups in a pyramidal organization appears to be one way to resolve this contradiction, allowing simultaneously for independence and coordination (Cernea 1985). The most successful models of nesting come from irrigation systems serving thousands of people at a time (McKean and Ostrom 1995). It is not clear whether such high levels of social organization are necessary or feasible to achieve forest health.

8. Conclusions

If forest health is an emerging commons, every new enclosure of the commons involves the infringement of somebody's personal liberty (Hardin 1968). Infringements made in the distant past are accepted because no contemporary complains of a loss. Newly proposed infringements to articulate monitoring and management responsibilities may be vigorously opposed by NIPF owners as violating property rights. But what do property rights mean? When landowners mutually agree to prevent and limit losses from natural threats, all forest owners become more free and perhaps more wealthy. As Hardin (1968) concludes by citing Hegel, "Freedom is the recognition of necessity"; individuals locked into the logic of the commons are free only to bring on universal ruin. Once they see the necessity of mutual coercion, they become free to pursue other goals.

Like individual parcellation, the recognition of common property gives resource owners the incentive to prevent and control insect damage, to make investments in forest health and to manage them sustainably and thus efficiently over the long term (McKean and Ostrom 1995). Forest health cannot be privately owned; it is an open-access resource. However, unlike individual parcellation, common property offers a way to continue productive use of the private aspects of a resource system while solving the monitoring and enforcement problems posed by the need to survey forest lands for insect problems.

9. References

Barry, T. 1987. The development of the hierarchy of effects: an historical perspective. Current Issues and Research in Advertising 54: 251-295.

Belanger, R.P., Hedden, R.L. and P.L. Lorio, Jr. 1993. Management strategies to reduce losses from the southern pine beetle. Southern Journal of Applied Forestry 17(3): 150-154.

Best, C., and L. A. Wayburn. 2001. America's Private Forests: Status and Stewardship. Washington, DC: Island Press.

Billings, R.F. and H.A. Pase, III. 1979. A Field Guide for Ground Checking SPB Spots. USDA Forest Service, Combined Forest Pest Research Development Program. Handbook No. 558. 19 p.

Birch, T. W. 1996. Private Forest Landowners of the United States, 1994. Proceedings of the Symposium on Non-Industrial Private Forests, Washington, D.C, pp. 10-18.

Birch, T. W. 1997. Private Forestland Owners of the Southern United States, 1994. Resource. Bull. NE-138. Radnor, PA: USDA Forest Service Northeastern Forest Experiment Station.

Blaikie, P. and H. Brookfield (eds.). 1987. Land Degradation and Society. London, UK: Methuen,

Bliss, J.C. and A.J. Martin. 1989. Identifying NIPF management motivations with qualitative methods. Forest Science 35(2): 601-622.

Bush, G.W. 2002. Healthy Forests: An Initiative for Wildfire Prevention and Stronger Communities. August 22, 2002. Washington DC: Office of the White House. Available at: http://www.whitehouse.gov/infocus/healthyforests/toc.html

Carlsson, Lars. 1996. The Swedish common forests: a common property resource in an urban, industrialized society. Rural Development Forestry Network Paper 20e Winter 1996/97: 1-14. Available at:
http://www.odi.org.uk/fpeg/publications/rdfn/20/rdfn-20e-i.pdf

Cernea, M.M. 1985. Alternative units of social organization sustaining afforestation strategies. Pp. 267-292, in Cernea, M.M. (ed.), Putting People First: Sociological Variables in Rural Development. New York: Oxford University Press.

Clawson, M. 1977. The economics of US nonindustrial private forests. Research Paper R-14. Washington, DC: Resources for the Future.

Dedrick, J. P., J. E. Johnson, T. E. Hall, and R. B. Hull. 1998. Attitudes of nonindustrial private forest landowners to ecosystem management in the United States: a review. Presented at Third IUFRO Extension Working Party Symposium, Extension Forestry: Bridging the Gap Between Research and Application. July 19-24, 1998, Blacksburg, Virginia, USA. Available at:
http://iufro.boku.ac.at/iufro/iufronet/d6/wu60603/proc1998/dedrick.htm

Dietz, T. Ostrom, E. N. Dolsak, P. Stern, S. Stonich, and E. Weber. 2001. Drama of the commons. Pp. 3-35 in The Drama of the Commons, Ostrom, E. T. Dietz, N. Dolsak, P. Stern, S. Stonich, and E. Weber, eds. National Research Council,

Dolsak, N. and E. Ostrom (editors). 2003. The Commons in the New Millennium: Challenges and Adaptation. Boston: The MIT Press.

Egan, A.F, and S.B. Jones. 1993. Do landowner practices reflect beliefs? Implications of an extension-research partnership. Journal of Forestry 91 (10): 39-45.

Ostrom, E. T. Dietz, N. Dolsak, P. Stern, S. Stonich, and E. Weber (editors). 2001. The Drama of The Commons. Washington D.C.: National Academy of Sciences Press.

Ervin, J., K. Larson, M. Miller, M. Washburn, and M. Webb. 2001. Nonindustrial private forest landowners: building the business case for sustainable forestry. Available at: http://sfp.cas.psu.edu/nipf.htm

Fecso, R.S., H.F. Kaiser, J.P. Royer and M. Weidenhammer. 1982. Management practices and reforestation decisions for harvested southern pinelands. Staff Rept. AGE5821230. Washington DC: USDA Statistical Reporting Service. 74 pp.

Feeny, D., F. Berkes, B.J. McCay, and J. M. Acheson. 1990. The tragedy of the commons: twenty-two years later. Human Ecology 18 (1): 1-19.

Flora, C. B. 2000. Measuring the social dimensions of managing natural resources. In Human Dimensions of Natural Resources Management: Emerging Issues and Practical Applications, eds. Fulton, D. C., K. C. Nelson, D. H. Anderson, and D. W. Lime. St. Paul: Cooperative Park Studies Program, University of Minnesota, Department of Forest Resources.

Geores, M. 2003. The relationship between resource definition and scale: considering the forest. Chapter in: N. Dolsak and E. Ostrom, Editors, The Commons in the New Millennium: Challenges and Adaptation. Cambridge: MIT Press.

Gibson, C., and C. D. Becker. 2000. A lack of institutional demand: why a strong local community in Western Ecuador fails to protect its forest. Pp. 135–161 in C. Gibson, M. A. McKean, and E. Ostrom, eds., People and Forests: Communities, Institutions, and Governance, ed. Cambridge, Mass.: MIT Press.

Hardin, G. 1968. The tragedy of the commons. Science, 162:1243-1248

Higman, S. Bass, S. Judd, N. Mayers, J. Nassbaum, R. 1999. The Sustainable Forestry Handbook. London: United Kingdom Limited and International Institute for Environment and Development.

Jones, S. B., A.E. Luloff, and J.C. Finley. 1995. Another look at NIPFs, facing our 'myths'. Journal of Forestry 93: 41-44.

Krogman, N., and T. Beckley. 2002. Corporate 'bail-outs' and local 'buyouts': pathways to community forestry? Society and Natural Resources 15:109-128.

Leopold, A. 1949. A Sand County Almanac. Oxford, UK: Oxford University Press.

McDonald, K. and R. McLain. 2003. The integration of community well-being and forest health in the Pacific Northwest. In Forest Communities, Community Forests. Kusel, J. Adler, E. Rowman & Littlefield Publishers. Inc.

McKean M. and E. Ostrom.1995. Common property regimes in the forest: just a relic from the past? Unasylva 180 (1): 3-21. Available at: http://www.fao.org/docrep/v3960e/v3960e03.htm#common%20property%20regimes%20in%20the%20forest:%20just%20a%20relic%20from%20the%20past

McKean M. A. 2000. Community governance of common property resources. Paper presented at the panel on "Governance and Civil Society," at the Fifth Annual Colloquium on Environmental Law and Institutions, "Sustainable Governance,"27-28 April 2000, Regal University Inn, Durham, North Carolina. Available at: http://www.law.duke.edu/news/papers/McKean2000.pdf

McKean, M. 1992. Success on the Commons: A Comparative Examination of Institutions for Common Property Resource Management. Journal of Theoretical Politics 4:3, July 247-281

McKean, M. 1996. Common property regimes as a solution to problems of scale and linkage. Pp. 223-243 in Susan Hanna, Carl Folke, and Karl-Göran Mäler, editors, Rights to Nature. Washington DC: Island Press.

McKean, M. 2000. Common property: what is it, what is it good for, and what makes it work? Chapter 2 in Keeping the Forest: Communities, Institutions, and the Governance of Forests. Clark Gibson and Elinor Ostrom, editors. Cambridge: MIT Press.

Meeker, J. R. W. N. Dixon, and J. L. Foltz. 1995. The Southern Pine Beetle, Dendroctonus frontalis Zimmermann. (Coleoptera: Scolytidae). Entomology Circular No. 369. Division of Plant Industry, Florida Department of Agriculture and Consumer Services. Available at: http://www.fl-dof.com/Pubs/pests/spb/spb.html

Merlo, M. 1995. Common property forest management in northern Italy: a historical and socio-economic profile. Unasylva 180 (1): 93-121. Available at: http://www.fao.org/docrep/v3960e/v3960e0a.htm#TopOfPage

Messerschmidt, D. A. 1993. Common forest resource management: annotated bibliography of Asia, Africa and Latin America. Rome: Food And Agriculture Organization of the United Nations. Available at: http://www.fao.org/DOCREP/006/U9040E/U9040E00.HTM#Contents

Molnar, J., J. Schelhas, and C. Holeski. 2003. Controlling the Southern Pine Beetle: Small Landowner Perceptions and Practices. Bulletin 649. Auburn: Alabama Agricultural Experiment Station, Auburn University.

Oakerson, R. 1986. A model for the analysis of common property problems. Pp.1330 in Proceedings of the Conference on Common Property Resource Management. National Academy Press, Washington DC, USA.

Ostrom, E. 1999. Self-governance and forest resources. Occasional paper No. 20. Center for International Forestry Research.

Ostrom, E., R. Gardner, and J. Walker. 1994. Rules, Games, and Common-Pool Resources. Ann Arbor: University of Michigan Press.

Peluso, N.L., C. Humphrey, and L. P. Fortmann. 1994. The rock, the beach, and the tidal pool: people and poverty in natural resource-dependent areas of the United States. Society and Natural Resources 7:1:23-38.

Price, T.S., C. Doggett, J.L. Pye and T.P. Holmes, eds. 1992. A history of SPB outbreaks in the southeastern United States. Sponsored by the Southern Forest Insect Work Conference. The Georgia Forestry Commission, Macon, GA. 65 p.

Richard, T., and E. Stein. 2003. Kicking dirt together in Colorado: community-ecosystem stewardship and the ponderosa pine forest partnership. In Forest Communities, Community Forests. Kusel, J. Adler, E. Rowman & Littlefield Publishers. Inc.

Schelhas, J., R. Zabawa, and J. Molnar. 2004. New opportunities for social research on forest landowners in the South. Southern Rural Sociology *In press.*

Schelhas, J. 2003. Race, Ethnicity, and Natural Resources in the U.S.: A Review. Natural Resources Journal 42(4): 723-763.

Stern, P. C., O. R. Young, and D. Druckman. 1992. Global Environmental Change: Understanding the Human Dimensions. Washington, D.C.: National Academy Press.

Thatcher, R.C. and P.J. Barry. 1982. Southern pine beetle. USDA Forest Service, Washington, D.C. Forest and Disease Leaflet No. 49. 7 p.

Ward J. D. and P. A. Mistretta. 2002. Impact of pests on forest health. In: Southern Forest Resource Assessment, edited by David N. Wear and John G. Greis, pp 403-428. General Technical Report SRS-53. Asheville, NC: USDA Forest Service, Southern Research Station

Williston, H. L., W. E. Balmer, and D.Tomczak. 1998. Managing the Family Forest in the South. Report SA-GR 22. Atlanta: USDA Forest Service.

Witzel, M. 2002. Management A-to-Z: AIDA. Web Site Dictionary of Business and Management. London: Financial Times. Available at: http://www.ftmastering.com/mmo/mmo02_3.htm

8

Evaluating Abiotic Factors Related to Forest Diseases: Tool for Sustainable Forest Management

Ludmila La Manna
Centro de Investigación y Extensión Forestal Andino Patagónico
Universidad Nacional de la Patagonia San Juan Bosco, CONICET
Argentina

1. Introduction

The influence of abiotic factors in the development of a disease is recognized in plant pathology. An abiotic factor may be the direct cause of a disease or may determinate the importance of an infectious disease or may be a key factor in forest decline diseases. Numerous studies have related forest diseases with abiotic factors around the world and for different forest species (Baccalla et al., 1998; Bernier & Lewis, 1999; Demchick & Sharpe, 2000; Dezzeo et al., 1997; Hennon et al., 1990; Horsley et al., 2000; Maciaszek, 1996).

Statistical techniques coupled with geographical information systems have fostered the development of predictive host habitat distribution models. The habitat-association approach can be used to generate risk maps, an important tool for developing forest management criteria (Fernández & Solla, 2006; Meentemeyer et al., 2004; Van Staden et al., 2004; Venette & Cohen, 2006). Many techniques with varying complexity were developed: rule based habitat models (Schadt et al., 2002a), niche modeling (Meentmayer et al., 2008, Rotemberry et al. 2006), neutral landscape models (With, 1997; With & King, 1997), etc.

This chapter aimed to describe some usefully methods for evaluating abiotic factors in relation to forest diseases at landscape scale and for developing risk models as tool for forest management. The methods described in this chapter were used for modeling *Phytophthora* disease risk in *Austrocedrus chilensis* [(D. Don) Pic. Serm. & Bizzarri] forests of Patagonia (La Manna et al., 2008b, 2012).

2. Collecting information

The predictive ability of a risk model is strongly associated with the quality and the level of detail of the habitat information on which the model is based. Developments and sophistication in remote sensing and geographical information systems have resulted in the potential for great increases in both the quality and quantity of habitat-level information that can be obtained and analyzed. These improved techniques also assist the study of forest pathology (Lundquist & Hamelin, 2005).

For developing risk models two issues are needed: a distributional map of the forest species and its health condition, in order to limit the study to the area of interest, and site thematic layers which were considered a priori as relevant for the disease occurrence.

The distribution map of the tree species and health status can be accomplished through techniques of varying complexity and cost. Currently, there are a variety of satellite images of different spatial, spectral and temporal resolution that can be applied to the study of forest ecosystems (Coppin et al., 2004; Iverson et al., 1989). The accuracy of the map will depend on the sensor's ability to discriminate the focal species from others, and its health status, based on measuring changes in electromagnetic energy (Karszenbaum, 1998).

Some of the sensors used for forest studies include Landsat Thematic Mapper (TM), Enhanced Thematic Mapper (ETM) and Multispectral Scanner (MSS), SPOT HRV, the Advanced Very High Resolution Radiometer (AVHRR), Advanced Spaceborne Thermal Emission and Reflection Radiometer (ASTER), QuickBird and Ikonos, at varying degrees of success (Chuvieco & Congalton, 1989; Franklin, 1994; Hyyppä et al., 2000; Lefsky et al., 2001; Martin et al., 1998; Peña & Altmann, 2009; Zhu & Evans, 1994). Sometimes, an intensive checking and corrections on the basis of field information are needed, and an iterative approach between image processing and field check must be applied (La Manna et al., 2008a). Aerial photographs are greatly useful for mapping and monitoring forests (Hennon et al., 1990; Holmström et al., 2001; Tuominen & Pekkarinen, 2005), however in many countries they are too expensive to acquire. It is also important to define if visual damage is enough for diagnosing the disease.

The site variables that should be included a priori in a risk model depend on the forest disease. Variables should be pre-selected based on current knowledge of the disease. For example, for mapping the risk of sudden oak death caused by *Phytophthora ramorum*, temperature and moisture variables were considered taking in account the pathogen persistence (Meentemeyer et al., 2004). For some species and areas of study, the wind was a relevant factor (Gardiner & Quine, 2000). For other forest diseases, the nutrition and soil characteristics were determinant factors (Bernier & Lewis, 1999; Demchik & Sharpe, 2000; Dezzeo et al., 1997; Horsley et al., 2000; Thomas & Büttner, 1998). *Austrocedrus chilensis* disease was associated with wet soils (La Manna & Rajchenberg, 2004a,b), agreeing with other diseases caused by *Phytophthora* species (Jönsson et al., 2005; Jung and Blaschke, 2004; Jung et al., 2000; Rhoades et al., 2003). Basing on this previous knowledge, climatic, topographic and edaphic thematic layers were considered for building the disease risk model. The environmental variables included in that case were mean annual precipitation, elevation, slope, aspect, distance to streams, and soil pH NaF (as indicator of allophane presence in volcanic soils) (La Manna et al., 2012).

On the other hand, the availability of information is also necessary taken into account. The quality and accuracy of the thematic layers will be the key for developing an useful risk model and for determining its scale. In this sense, there is great disparity in the information available according to the country or the region of study (Matteucci, 2007). However, the access to free information has greatly increased in recent years. For example, Google Earth (www.earth.google.com) may be a good tool for characterizing geomorphologies and drainage systems. Digital elevation models are also freely available. The Global Digital Elevation Model (GDEM) from ASTER has 30m resolution and covers the 99% of the earth surface. The Shuttle Radar Topography Mission (SRTM) obtains altitude data by radar interferometry and covers the 80% of the earth surface. This sensor has 90m resolution, and it also has a 30m resolution band with a lower coverage. Both elevation digital models present advantages and disadvantages (Hayakawa et al., 2008).

Global climate data can be freely obtained from the global grid of precipitation (www.worldclim.org), with 1km spatial resolution (Hijmans et al., 2005). Sometimes, local weather stations or local climate models may provide more useful data since they have a greater level of detail. The greater resolution data may significantly improve risk assessment (Krist et al., 2008).

3. Developing the database

Developing risk models require both, forest health condition and abiotic factors, to be combined in a geographic information system (GIS). In this chapter, tools from Arcview 3.3 and ERDAS software are described, but newer software for editing GIS have similar tools. On the other hand, researchers around the world are developing free GIS, which now or in the future will probably have the same tools.

The database must include information from training sites, i.e. geo-located forests patches whose health condition and abiotic factors are known. Training sites can be selected from field checking (La Manna et al., 2012) or from the map of species distribution and health condition (La Manna et al., 2008b). The patches should have an homogeneous health conditions; and training sites should include diseased and healthy patches or just diseased ones, depending on model requirements. The selection of training sites requires a proper sampling method, covering the range of host and abiotic conditions in order to minimize bias. A stratified-random sampling or a random sampling should be applied, and the extension Table Select deluxe tools v.1.0 of Arc View software can be useful for selection.

The abiotic factors should be mapped in all the study area. Environmental features of training sites are needed to build the database; but the environmental features along all the area of distribution of the forest species are needed to build the risk map. Figure 1 schematizes the process for building database.

Once the environmental layers are complete, the mean values of each site attributes layer can be extracted by the Zonal attributes tool of ERDAS software for each training site. This tool enables to extract the zonal statistics (mean, standard deviation, minimum and maximum) from a vector coverage and save them as polygon attributes.

4. Building risk models

There are different modeling techniques for developing risk model based on abiotic factors, with predictive performance varying according to the focus of the study (Brotons et al., 2004; Manel et al., 2001; Pearson et al., 2006, 2007; Phillips et al. 2006). Data requirements vary between the techniques. While some models require data of presence and absence of the disease (i.e., diseased and healthy training sites), others need only presence data. The former models are appropriate if absence of the disease is due to environmental restrictions, while the latter approach is appropriate when factors other than environmental variables (e.g. history of spread) explain most of the absences.

In some cases, absence data are doubtful; for example for forest diseases that are manifested earlier in the lower stem and latter in the crown, delaying detection by remote sensing. In these cases, health condition of training sites should be obtained from the field (La Manna et al., 2012), since failure to detect absences results in false negatives, which change mathematical functions describing habitats.

Among the available modeling techniques, three are described in this chapter on the basis of their requirements on disease presence or disease presence/absence data: Mahalanobis distance (requires only presence data), Maxent (requires only presence data and generates

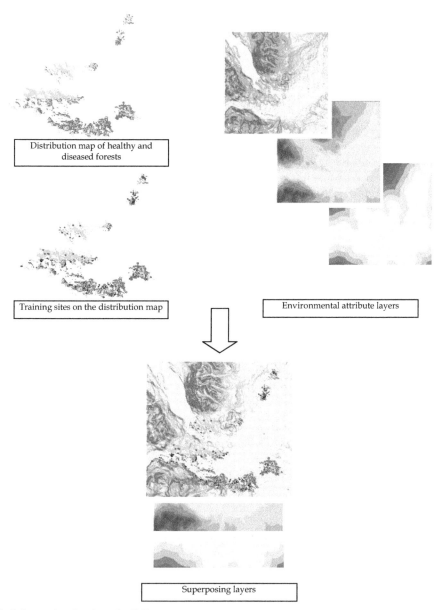

Fig. 1. Scheme for database building

pseudo-absences) and Logistic regression (based on presence/absence data). These methods are inherently flexible, being applicable to a wide range of ecological questions, taxonomic units, and sampling protocols and they produced useful predictions in other studies (DeVries, 2005; Elith et al., 2006; Hellgren et al., 2007; La Manna et al., 2008b, 2012; Marsden & Fielding, 1999; Pearson et al., 2006;Schadt et al., 2002b).

4.1 Mahalanobis distance model
4.1.1 Brief description of the mathematical model

Mahalanobis distance, which requires only presence records, projects the potential distribution of the disease into a geographical space without giving weight to observed absence information (Pearson et al., 2006). Mahalanobis distance was introduced by Mahalanobis (1936) and it is the standardized difference between the values of a set of environmental variables describing a site (rasterized cell or pixel in a GIS) and the mean values for those same variables calculated from points at which the disease was detected (Browning et al., 2005; Rotenberry et al., 2006). Mahalanobis distances are based on both the mean and variance of the predictor variables, plus the covariance matrix of all the variables. Mahalanobis distance is calculated as:

$$D^2=(x-m)^T C^{-1} (x-m) \tag{1}$$

where:
D^2 =Mahalanobis distance
x=Vector of data
m=Vector of mean values of independent variables
C^{-1}= Inverse Covariance matrix of independent variables
T=Indicates vector should be transposed

The greater the similarity of environment conditions in a point with mean environmental conditions in all training points, the smaller the Mahalanobis distance and the higher the disease risk at that point. Mahalanobis distance has been used in studies employing a GIS to quantify habitat suitability for wildlife and plant species (DeVries, 2005; Johnson & Gillingham, 2005; Hellgren et al., 2007).

4.1.2 Applying Mahalanabis distance in a GIS

Since Mahalanobis distance considers points (and not patches), the polygon layer with the diseased training sites, selected in the field or from the map, must be converted to a point layer. This conversion is done founding the point at the center of each patch, by "Convert shape to centroid" option from Xtool ArcView extension. The vector of mean values for each site variable and the variance/covariance matrix for site variables is generated from this point layer (Figure 2).

The Mahalanobis distance for each cell of the study area is calculated based on this matrix with Mahalanobis distances extension for ArcView (Jenness, 2003). This extension may be freely downloaded from: http://www.jennessent.com/arcview/mahalanobis.htm. For an easier interpretation of results, the Mahalanobis distance statistic can be converted to probability values rescaling to range from 0 to 1 according to χ^2 distribution (Rotenberry et al., 2006).

4.2 Maximum entropy species distribution modelling (Maxent)
4.2.1 Brief description of the mathematical model

Maxent, as Mahalanobis distance, is a model requiring presence data, but it generates "pseudo-absences" using background data as substitute for true absences (Phillips and Dudík, 2008). Thus, Maxent formalizes the principle that the estimated distribution must agree with everything that is known (or inferred from the environmental conditions at the occurrence localities) but should avoid placing any unfounded constraints. The approach is

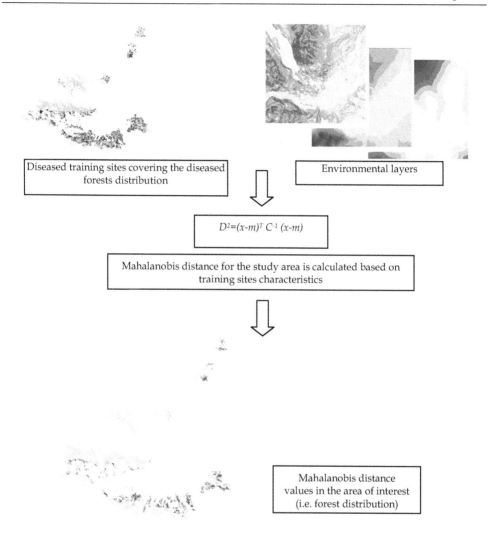

Fig. 2. Schematic representation of Mahalanobis distance procedure

to find the probability distribution of maximum entropy (i.e, closest to uniform, or most spread out), subject to constraints imposed by the information available regarding the observed distribution of the disease and environmental conditions across the study area. The Maxent distribution belongs to the family of Gibb's distributions and maximizes a penalized log likelihood of the presence sites. The mathematical definition of Maxent and the detailed algorithms are described by Phillips et al. (2006), Phillips & Dudík (2008) and Elith et al. (2011).

Maxent has been applied to modeling species distributions and disease risk with good performance (La Manna et al., 2012; Pearson et al., 2007; Phillips & Dudík, 2008; Phillips et al., 2006).

4.2.2 Applying Maxent in a GIS

Maxent can be freely downloaded and used from: http://www.cs.princeton.edu/~schapire/maxent/ and it is regularly updated to include new capabilities. A friendly tutorial explaining how to use this software is provided in the web page, including a Spanish translation.

To perform a run, a file containing presence localities (i.e. diseased training sites), and a directory containing environmental variables need to be supplied. The implementation of Maxent requires the conversion of the files to proper formats. The file with the list of diseased training sites must be in csv format, including their identification name, longitude and latitude. The environmental layers must be saved as ascii raster grids (i.e. .asc format) and the grids must all have the same geographic bounds and cell size. Environmental grids can be saved as ascii file by "Export data source" tool of ArcView. Maxent must be run following the detailed information included in the tutorial (Phillips et al., 2005).

Maxent supports three output formats for model values: the Maxent exponential model itself (raw), cumulative and logistic. The logistic output format, with values between 0 and 1, is easier interpreted and it improves model calibration, so that large differences in output values correspond better to large differences in suitability (Phillips & Dudík, 2008).

4.3 Logistic regression model
4.3.1 Brief description of the mathematical model

The logistic regression is a generalized linear model used for binomial regression, and requires presence/absence data. What distinguishes a logistic regression model from the linear regression model is that the dependent variable is binary or dichotomous (Hosmer & Lemeshow, 1989). The binary dependent variable is disease occurrence (i.e., diseased training site; y=1) and disease absence (i.e., healthy training site; y=0). In contrast to others described models, Logistic regression projects the potential distribution of the disease onto a geographical space whereby information regarding unsuitable conditions resulting from environmental constraints is inherent within the absence data (Pearson et al., 2006).

Logistic regression predicts the probability of occurrence of an event by fitting data to a logistic curve, presenting the following formula:

$$\text{Logit } (P) = \beta_0 + \beta_1 \times V1 + \beta_2 \times V2 + \ldots + \beta_n \times Vn \tag{2}$$

where P is the probability of disease occurrence

β_0 is the Y-intercept

$\beta_1 \ldots \beta_n$ are the coefficients assigned to each of the independent variables (V1... Vn)

Probability values are calculated based on the equation below, where e is the natural exponent:

$$P = e^{\text{logit}(P)} / 1 + e^{\text{logit}(P)} \tag{3}$$

A comprehensive description of logistic regression and its applications is presented by Hosmer & Lemeshow (1989). Figure 3 shows a graphical example of a logistic regression model based on presence/absence data of a disease and a soil feature as independent variable.

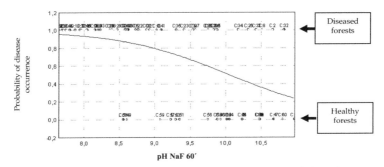

Fig. 3. Observed and estimated (by logistic regression model) probability of *Austrocedrus chilensis* disease according to soil pH NaF values.

4.3.2 Applying logistic regression in a GIS

From the database combining health condition and abiotic factors from training sites, the logistic regression model can be performed using common statistical software, as SPSS, SAS, Infostat, or free software packages. For example, Infostat is a friendly and economic statistical software and it offers a version that can be freely downloaded from: http://www.infostat.com.ar.

The output of logistic regression analysis shows the coefficients assigned to each of the environmental variables (V1... Vn), and the probabilities values for each cell of the study area can be obtained in the GIS. Calculations can be done with "Calculate maps" tool from Grid Analyst extension of ArcView, considering site layers in grid format (Figure 4). Thus, a grid with probabilities of disease occurrence is generated according to the logistic model.

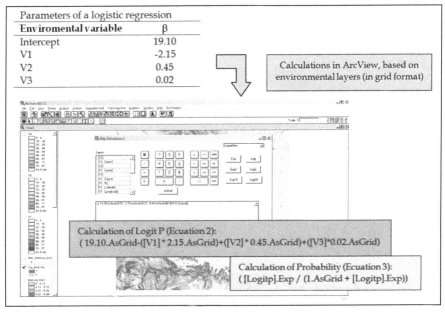

Fig. 4. Example of logistic regression model applied in a GIS.

4.4 Evaluating abiotic factors selecting the most important variables

An advantage of Maxent and the logistic regression models respect to Mahalanobis distance, is that the former allow easily discriminating the abiotic factors most related to the disease and choosing the better combination of variables. As mentioned above, environmental variables included a priori in the models depend on the knowledge about the disease. However, not all the variables considered a priori could be equally important for quantifying the disease risk at the landscape scale.

Maxent allows detecting which variables matter most, calculating the percent contribution to the model for each environmental variable (Phillips et al., 2005). As alternative estimates of variable´s weight, a jackknife test can also be run by Maxent. Figure 5 shows an example of jackknife test, where the environmental variable "agua-move" appears to have the most useful information by itself (blue bar). The environmental variable that decreases the gain the most when it is omitted is also agua_move (light blue bar), which therefore appears to have the most information that is not present in the other variables.

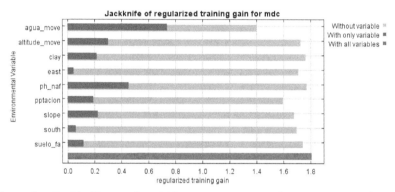

Fig. 5. Example of jackknife test of variable importance according to Maxent software.

In the case of logistic regression the better combination of variables can be chosen according to the best subsets selection technique (Hosmer and Lemeshow, 1989), the lowest Akaike information criterion (AIC) (Burnham & Anderson, 1998), the greatest sensitivity (i.e., proportion of correctly predicted disease occurrences) or the stepwise method (Steyerberg et al., 1999).

4.5 Assessment of model performance

The predictive performance of modeling algorithms may be very different (Brotons et al., 2004; Manel et al., 2001; Pearson et al., 2006, 2007; Phillips et al., 2006). Differences could be related to the intrinsic properties of mathematical functions inherent to each model and to the various assumptions made by each algorithm when extrapolating environmental variables beyond the range of the data used to define the model (Pearson et al., 2006). Further, the set of data for running the models differs according to consider presence or presence/absence data.

Receiver operating characteristic (ROC) curves and Kappa statistic are index widely used for assessing performance of models. ROC curve procedure is a useful way to evaluate the performance of classification schemes in which there is one variable with two categories by which subjects are classified. The area under the ROC curve (AUC) is the probability of a randomly chosen presence site being ranked above a randomly chosen absence site. This procedure relates relative proportions of correctly and incorrectly classified predictions over a wide and continuous range of threshold levels (Pearce & Ferrier, 2000). The main advantage of this analysis is that AUC provides a single measure of model performance, independent of any particular choice of threshold. AUC can be calculated with common statistical software. ROC plot showed in Figure 6 is obtained by plotting all sensitivity values (true positive fraction) on the y axis against their equivalent (1−specificity) values (false positive fraction) on the x axis. Specificity of a model refers to the proportion of correctly predicted absences.

ROC analysis has been applied to a variety of ecological models (Brotons et al., 2004; Hernández et al., 2006; La Manna et al. 2008b, Pearson et al., 2006; Phillips et al., 2006). Values between 0.7 and 0.9 indicate a reasonable discrimination ability considered potentially useful, and rates higher that 0.9 indicate very good discrimination (Swets, 1988). If absence data are not available, AUC may also be calculated with presence data and pseudo-absences chosen uniformly at random from the study area (Phillips et al., 2006). However, counting with both true absence and presence sites is better for evaluating model performance (Fielding & Bell, 1997).

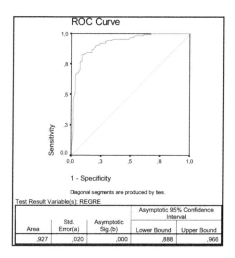

Fig. 6. Example of ROC curve obtained for a regression model by SPSS software.

Kappa statistic is another index widely used (Loiselle et al., 2003; Hérnández et al., 2006; Pearson et al., 2006), that can be calculated with common statistical software. The Cohen's Kappa and Classification Table Metrics 2.1a, an ArcView 3x extension, may also be useful and can be freely downloaded from: http://www.jennessent.com/arcview/kappa_stats.htm. Cohen's kappa is calculated at thresholds increments, e.g. increments of 0.05, from 0 to 1, and the maximum Kappa for each model is considered. Kappa values approaching 0.6 represent a good model (Fielding & Bell, 1997).

The models should be run on the full set of training data, to provide best estimates of the disease's potential distribution (Philips et al., 2006). However, in order to assess and to compare the model performance, models should be run with just a portion of the training sites and the rest of data should be used for the assessment. For each model, some (e.g. ten) random partitions of data are done maintaining the remaining 25% of training sites for performance assessment. Then, AUC and Kappa values are calculated for each random set of assessment data and for each model, and they are compared between models by non-parametric analysis (Philips et al., 2006).

The performance of the three models described in this chapter (i.e. Mahalanobis distance, Maxent and Logistic Regression) was compared for modeling a forest disease in Patagonia (La Manna et al., 2012). Results showed that all the models were consistent in their prediction; however, Maxent and Logistic regression presented a better performance, with greater values of AUC and Kappa statistics; and logistic regression allowed the best discrimination of high risk sites. Studies that compared presence-absence versus presence-only modeling methods, suggest that if absence data are available, methods using this information should be preferably used in most situations (Brotons et al., 2004). However, Maxent is considered as one of the best performing models (Elith et al. 2006; Hernández et al., 2006; Pearson et al., 2006; Phillips et al., 2006), and Mahalanobis distance also provided good results in conservation studies (DeVries, 2005; Johnson & Gillingham, 2005; Hellgren et al., 2007).

The performance of the risk models may greatly vary in each case and forest disease. Building and comparing models based on different algorithms allow finding the best.

4.6 Mapping the risk. Selecting thresholds

The three risk models presented in this chapter have as result grids with probabilities values of disease occurrence, varying between 0 and 1. However, for proposing management criteria is important to define what probability represents a high risk of disease. 0.4?, 0.5?, 0.7?... In order to convert quantitative measures of disease risk (i.e., probability) to qualitative values (i.e., low, moderate or high risk) threshold values must be selected.

A possible criterion is to define thresholds by maximizing agreement between observed and modeled distributions for the sampled dataset. Sensitivity (the proportion of true positive predictions vs. the number of actual positive sites) and specificity (the proportion of true negative predictions vs. the number of actual negative sites) are calculated at different thresholds according to AUC coordinates. The threshold at which these two values are closest can be adopted. This approach balances the cost arising from an incorrect prediction against the benefit gained from a correct prediction (Manel et al., 2001), and is one of the recommended criteria for selecting thresholds (Liu et al., 2005).

The lowest predicted value associated with any one of the observed presence records can also considered as a threshold (i.e, lowest presence threshold) (Pearson et al., 2007). This approach can be interpreted ecologically as identifying pixels predicted as being at least as suitable as those where the disease presence has been recorded. The threshold identifies the maximum predicted area possible whilst maintaining zero omission error in the training data set.

Using the two thresholds, three risk categories can be defined: low (with p values lower than the lowest presence threshold); moderate (p values between the lowest and the sensitivity-specificity approach thresholds); and high risk (p values greater than the sensitivity-specificity approach threshold) (La Manna et al. 2012).

Risk maps of disease occurrence can be generated for each model by reclassifying the model outputs, using Grid analyst extension of ArcView software.

5. Conclusions

Forest diseases are key determinants of forest health, and information about disease presence and potential distribution are important to any management decision. Risk maps are more likely to be used if they addresses the same scale at which management decisions are made. Stand scale management is increasingly being supplemented or replaced by landscape-scale management (Lundquist, 2005). Forest diseases risk assessment provides important information to the forest services that makes critical decisions on the best allocation of often-scarce resources.

Risk models for pine wilt disease (*Bursaphelenchus xylophilus*) in Spain allowed planning control actions and preventing to plant susceptible species in the high risk areas (Fernández & Solla, 2006). Risk models for sudden oak death in California provide an effective management tool for identifying emergent infections before they become established (Meentemeyer et al., 2004). Risk models for economically important South African plantation pathogens allowed to asses the impact of climate change on the local forestry industry (Van Staden et al., 2004). Risk maps for *A. chilensis* disease in a valley of Patagonia allowed to detect healthy forests at risk only inside protected areas. These results allowed to suggest management actions for cattle and logging in disease-prone sites. This risk map also provided useful information for preventing restock in areas where the risk is greatest (La Manna et al., 2012).

Risk models discussed in this chapter allowed the evaluation of abiotic factors related to the disease. This kind of models provides important information, which can be improved if knowledge about the biology and spreading of a causal biotic agent is available. It is important to know whether the forest pathogen under study is endemic or exotic. If it is exotic, the susceptibility must be assessed, based on the biological availability of a host and the potential for introduction and establishment of the disease within a predefined time frame. For this evaluation, the connectivity between patches may be key (Ellis et al., 2010). On the other hand, if it is endemic, the disease is already established throughout a region, and then a susceptibility assessment is not required because the potential or source for actualized harm is assumed to be equal everywhere (Krist et al., 2006).

For both endemic and exotic diseases, mortality occurrence may vary greatly depending on site and stand conditions, and models like those shown in this chapter are a good tool for assessing risk. Variables included in the models should be carefully pre-selected according to the previous knowledge about the disease. These models (i.e., Mahalanobis distance, Maxent and Logistic Regression) also admit variables like distance to roads, or distance to foci of infection, that could be important for spreading of infectious diseases.

6. Acknowledgment

The publication of this chapter was funded by Universidad Nacional de la Patagonia San Juan Bosco (PI 773).

7. References

Baccalá, N.; Rosso, P. & Havrylenko, M. (1998). *Austrocedrus chilensis* mortality in the Nahuel Huapi National Park (Argentina). Forest Ecology and Management , Vol. 109, pp. 261-269, ISSN 0378-1127

Bernier, D. & Lewis, K. (1999). Site and soil characteristics related to the incidence of *Inonotus tomentosus*. Forest Ecology and Management , Vol. 120, pp. 131-142, ISSN 0378-1127

Brotons, L.; Thuiller, W.; Araújo, M. & Hirzel, A. (2004). Presence-absence versus presence-only modelling methods for predicting bird habitat suitability. Ecography, Vol. 27, pp. 437-448, ISSN 0906-7590

Browning, D.M.; Beaupré, S.J. & Duncan, L. (2005). Using partitioned Mahalanobis D2(k) to formulate a GIS-based model of timber rattlesnake hibernacula. Journal of Wildlife Management, Vol. 69, No. 1, pp. 33-44, ISSN 1937-2817

Burnham, K.P. & Anderson, D.R. (1998). Model Selection and Inference: A Practical Information-Theoretic Approach. Springer Verlag, ISBN 0-387-95364-7, New York

Chuvieco, E. & Congalton, R. (1989). Application of remote sensing and geographic information systems to forest fire hazard mapping. Remote Sensing of Environment, Vol. 29, No. 2 (August 1989), pp. 147-159, ISSN 0034-4257

Coppin, P.; Jonckheere, I.; Nackaerts, K.; Muys, B. & Lambin, E. (2004). Digital change detection methods in ecosystem monitoring: a review. International Journal of Remote Sensing Vol. 25, No. 9, pp. 1565–1596, ISSN 0143-1161

Demchik, M.C. & Sharpe, W.E. (2000). The effect of soil nutrition, soil acidity and drought on northern red oak (*Quercus rubra* L.) growth and nutrition on Pennsylvania sites with high and low red oak mortality. Forest Ecology and Management, Vol. 136, pp. 199-207, ISSN 0378-1127

Dezzeo, N.; Hernández, L. & Fölster, H. (1997). Canopy dieback in lower montane forests of Alto Urimán, Venezuelan Guayana. Plant Ecology, Vol. 132, pp. 197-209, ISSN 1385-0237

DeVries, R.J. (2005). Spatial Modelling using the Mahalanobis Statistic: two examples from the discipline of Plant Geography. In Proceedings of the International Congress on Modelling and Simulation. Zerger, A. & Argent, R.M. Eds.), Modelling and Simulation Society of Australia and New Zealand.

Elith, J.; Graham, C.; Anderson, R.; Dudík, M.; Ferrier, S.; Guisan, A.; Hijmans, R.; Huettmann, F. et al. (2006). Novel methods improve prediction of species' distributions from occurrence data. Ecography, Vol. 29, pp. 129-151, ISSN 0906-7590

Elith, J.; Phillips, S.; Hastie, T.; Dudík, M.; En Chee, Y. & Yates, C. (2011). A statistical explanation of MaxEnt for ecologists. Diversity and Distributions, Vol. 17, pp. 43–57, ISSN 1472-4642

Ellis, A.M.; Václavík, T. & Meentemeyer, R.K. (2010). When is connectivity important? A case study of the spatial pattern of sudden oak death. Oikos, Vol. 119, No. 3, pp. 485–493, ISSN 0030-1299

Fernández, J.M. & Solla, A. (2006). Mapas de riesgo de aparición y desarrollo de la enfermedad del marchitamiento de los pinos (*Bursaphelenchus xylophilus*) en Extremadura. Investigación Agraria Sistemas y Recursos Forestales, Vol. 15, pp. 141–151, ISSN 1131-7965

Fielding, A.H. & Bell, J. (1997). A review of methods for the assessment of prediction errors in conservation presence/absence models. Environmental Conservation, Vol. 24, pp.38–49, ISSN 0376-8929

Franklin, S. E. (1994). Discrimination of subalpine forest species and canopy density using CASI, SPOT, PLA, and Landsat TM data. Photogrammetric Engineering and Remote Sensing , Vol. 60, pp. 1233–1241, ISSN 0099-1112

Gardiner, B. & Quine, C. (2000). Management of forests to reduce the risk of abiotic damage — a review with particular reference to the effects of strong winds. Forest Ecology and Management, Vol. 135, No.1-3, (September 2000), pp. 261-277, ISSN 0378-1127

Hayakawa, Y.; Oguchi, T. & Lin, Z. (2008). Comparison of new and existing global digital elevation models: ASTER G-DEM and SRTM-3. Geophysical Research Letters, Vol. 35, No. 17, pp. 1-5, ISSN 0094-8276

Hellgren, E.C.; Bales, S.L.; Gregory, M.S.; Leslie, D.M. & Clark, J.D. (2007). Testing a Mahalanobis distance model of black black bear habitat use in the Ouachita Mountains of Oklahoma. Journal of Wildlife Management, Vol. 71, No. 3, pp. 924-928, ISSN 1937-2817

Hennon, P.E.; Hansen, E.M. & Shaw III, C.G. 1990. Dynamics of decline and mortality in *Chamaecyparis nootkatensis* in southeast Alaska. Canadian Journal of Botany, Vol. 68, pp. 651-662, ISSN 0008-4026

Hernández, P.; Graham, C.; Master, L. & Albert, D. (2006). The effect of sample size and species characteristics on performance of different species distribution modelling methods. Ecography, Vol. 29, pp. 773-785, ISSN 0906-7590

Hijmans, R.J.; Cameron, S.; Parra, J.; Jones, P. & Jarvis, A. (2005). Very High Resolution Interpolated Climate Surfaces for Global Land Areas. International Journal of Climatology, Vol. 25, pp. 1965–1978, ISSN 0899-8418

Holmström, H.; Nilsson, M. & Ståhl, G. (2001). Simultaneous Estimations of Forest Parameters using Aerial Photograph Interpreted Data and the k Nearest Neighbour Method. Scandinavian Journal of Forest Research, Vol. 16, No. 1 (January 2001), pp. 67-78, ISSN 0282-7581

Horsley, S. B.; Long, R. P.; Bailey, S. W.; Hallet, R. & Hall, T. 2000. Factors associated with the decline disease of sugar maple on the Allegheny Plateau. Canadian Journal of Forest Research, Vol. 30, pp. 1365-1378, ISSN 1208-6037

Hosmer, D.W. & Lemeshow, S. (1989). Applied Logistic Regression. John Wiley & Sons, ISBN 0-471-61553-6, New York.

Hyyppä, J.; Hyyppä, H.; Inkinen, M.; Engdahl, M.; Linko, S. & Zhu, Y.H. (2000). Accuracy comparison of various remote sensing data sources in the retrieval of forest stand attributes. Forest Ecology and Management, Vol. 128, No. 1-2, (March 2000), pp. 109-120, ISSN 0378-1127

Iverson, L.R.; Graham R.L. & Cook, E.A. (1989). Applications of satellite remote sensing to forested ecosystems. Landscape Ecology, Vol. 3, No. 2, pp. 131-143, ISSN 1572-9761

Jenness, J. (2003). Mahalanobis distances (mahalanobis.avx) extension for ArcView 3.x, Jenness Enterprises. Available from: http://www.jennessent.com/arcview/mahalanobis.htm.

Johnson, C.J. & Gillingham, M.P. (2005). An evaluation of mapped species distribution models used for conservation planning. Environmental Conservation, Vol. 32, No. 2, pp. 1–12, ISSN 0376-8929

Jönsson, U.; Jung, T.; Sonesson, K. & Rosengren, U. (2005). Relationships between health of Quercus robur, occurrence of Phytophthora species and site conditions in southern Sweden. Plant Pathology, Vol. 54, pp. 502–511, ISSN 0032-0862

Jung, T. & Blaschke, M. (2004). Phytophthora root and collar rot of alders in Bavaria: distribution, modes of spread and possible management strategies. Plant Pathology, Vol. 53, pp. 197–208, ISSN 0032-0862

Jung, T.; Blaschke, H. & Oûwald, W. (2000). Involvement of soilborne Phytophthora species in Central European oak decline and the effect of site factors on the disease. Plant Pathology, Vol. 49, pp. 706-718, ISSN 0032-0862

Karszenbaum, H. (1998). Procesamiento de imágenes satelitales para la gestión ambiental, In: Sistemas ambientales complejos: herramientas de análisis espacial, Matteucci, S.D. & Buzai, G.D. (Eds.), pp. 197-217, Eudeba, ISBN 950-23-0760-7, Buenos Aires

Krist, F.; Sapio, F.& Tkacz, B. (2006). A Multi-Criteria Framework for Producing Local, Regional, and National Insect and Disease Risk Maps. USDA Forest Service. http://www.fs.fed.us/foresthealth/technology/pdfs/hazard-risk-mapmethods.pdf

La Manna, L. & Rajchenberg, M. (2004a). The decline of Austrocedrus chilensis forests in Patagonia, Argentina: soil features as predisposing factors. Forest Ecology and Management, Vol. 190, pp. 345-357, ISSN 0378-1127

La Manna, L. & Rajchenberg, M. (2004b). Soil properties and Austrocedrus chilensis decline in Central Patagonia, Argentina. Plant and Soil, Vol. 263, pp. 29-41, ISSN 0032-079X

La Manna, L.; Carabelli, F.; Gómez, M. & Matteucci, S.D. (2008a). Disposición espacial de parches de *Austrocedrus chilensis* con síntomas de defoliación y mortalidad en el Valle 16 de Octubre (Chubut, Argentina). Bosque, Vol. 29, No 1, pp. 23-32, ISSN 0717-9200

La Manna, L.; Mateucci, S.D. & Kitzberger, T. (2008b). Abiotic factors related to the incidence of *Austrocedrus chilensis* disease at a landscape scale. Forest Ecology and Management, Vol. 256, pp. 1087-1095, ISSN 0378-1127

La Manna, L.; Mateucci, S.D. & Kitzberger, T. (2012). Modelling Phythophtora disease risk in *Austrocedrus chilensis* forests of Patagonia. European Journal of Forest Research, Vol. 131, Issue 2, pp. 323-337, ISSN 1612-4669.

Lefsky, M.; Cohen, W. & Spies, T. (2001). An evaluation of alternate remote sensing products for forest inventory, monitoring, and mapping of Douglas-fir forests in western Oregon. Canadian Journal of Forest Research, Vol. 31, pp. 78–87, ISSN 1208-6037

Liu, C.; Berry, P.; Dawson, T. & Pearson, R. (2005). Selecting thresholds of occurrence in the prediction of species distributions. Ecography, Vol. 28, pp. 385-393, ISSN 0906-7590

Loiselle, B.; Howell, C.; Graham, C.; Goerck, J.; Brooks, T.; Smith, K. & Williams, P. (2003). Avoiding pitfalls of using species-distribution models in conservation planning. Conservation Biology, Vol. 17, No 6, pp. 1–10, ISSN 0888-8892

Lundquist, J.E. (2005). Landscape pathology – Forest pathology in the era of landscape ecology. In: Forest pathology: from genes to landscape, Lundquist, J.E. & Hamelin, R.C. (Eds.), pp. 155-165, APS Press, ISBN 0-89054-334-8, St. Paul, Minnesota

Lundquist, J.E. & Hamelin, R.C. (2005). Forest pathology: from genes to landscape. APS Press, ISBN 0-89054-334-8, St. Paul, Minnesota

Maciaszek, W. (1996). Pedological aspects of oak decline in south-eastern Poland. Prace Instytutu Badawczego Lésnictwa Vol. 824, pp. 89-109, ISSN 1732-9442

Mahalanobis, P.C. (1936). On the generalized distance in statistics. Proceedings of the National Institute of Sciences of India, Vol. 2, No 1, pp. 49–55. Available from: http://www.insa.ac.in/insa_pdf/20005b8c_49.pdf

Manel, S.; Williams, H.C.& Ormerod, S.J. (2001) Evaluating presences-absence models in ecology: the need to account for prevalence. Journal of Applied Ecology, Vol. 38, pp. 921–931, ISSN 0021-8901

Marsden, S. & Fielding, A. (1999). Habitat associations of parrots on the Wallacean islands of Buru, Seram and Sumba. Journal of Biogeography, Vol. 26, pp. 439–446, ISSN 0305-0270

Martin, M.; Newman, S.; Aber, J. & Congalton, R. (1998). Determining Forest Species Composition Using High Spectral Resolution Remote Sensing Data. Remote sensing of environment, Vol. 65, pp. 249–254, ISSN 0034-4257

Matteucci, S.D. (2007). Los Sin Dato. Una propuesta para pensar, mejorar y ejecutar. Fronteras, Vol. 6, No 6, pp. 41-44, ISSN 1667-3999

Meentemeyer, R.; Rizzo, D.; Mark, W. & Lotz, E. (2004). Mapping the risk of establishment and spread of sudden oak death in California. Forest Ecology and Management, Vol. 200, pp. 195-214, ISSN 0378-1127

Meentemeyer, R.K.; Anacker, B.; Mark, W & Rizzo, D. (2008). Early detection of emerging forest disease using dispersal estimation and ecological niche modeling. Ecological Applications, Vol. 18, pp. 377-390, ISSN 1051-0761

Pearce, J. & Ferrier, S. (2000). Evaluating the predictive performance of habitat models developed using logistic regression. Ecological Modelling, Vol. 133, pp. 225–245, ISSN 0304-3800

Pearson, R.; Raxworthy, C.; Nakamura, M. & Townsend, P. (2007). Predicting species distributions from small numbers of occurrence records: a test case using cryptic geckos in Madagascar. Journal of Biogeography, Vol. 34, pp. 102–117, ISSN 0305-0270

Pearson, R.; Thuiller, W.; Araujo, M.; Martinez-Meyer, E.; Brotons, L.; McClean, C.; Miles, L.; Segurado, P.; Dawson, T. & Lees, D. (2006). Model-based uncertainty in species range prediction. Journal of Biogeography, Vol. 33, pp. 1704–1711, ISSN 0305-0270

Peña, M.A. & Altmann, S. (2009). Use of satellite-derived hyperspectral indices to identify stress symptoms in an *Austrocedrus chilensis* forest infested by the aphid *Cinara cupressi*. International Journal of Pest Management, Vol. 55, No. 3, pp. 197-206, ISSN 0967-0874

Phillips, S. & Dudík, M. (2008). Modeling of species distributions with Maxent: new extensions and a comprehensive evaluation. Ecography, Vol. 31, pp. 161-175., ISSN 0906-7590

Phillips, S.; Anderson, R. & Schapired, R. (2005). Maxent software for species distribution modeling. Available from: http://www.cs.princeton.edu/_schapire/ maxent.

Phillips, S.; Anderson, R. & Schapired, R. (2006). Maximum entropy modeling of species geographic distributions. Ecological Modelling, Vol. 190, pp. 231–259, ISSN 0304-3800

Rhoades, C.; Brosi, S.; Dattilo, A.; Vincelli, P. (2003). Effect of soil compaction and moisture on incidence of *phytophthora* root rot on American chestnut (*Castanea dentata*) seedlings. Forest Ecology and Management, Vol. 184, pp. 47–54, ISSN 0378-1127

Rotenberry, J.T.; Preston, K.L. & Knick, S.T. (2006). GIS-based niche modeling for mapping species´ habitat. Ecology, Vol. 87, No 6, pp. 1458-1464, ISSN 0012-9658

Schadt, S.; Knauers, F.; Kaczensky, P.; Revilla, E.; Wiegand, T. & Trepl, L. (2002a). Rule-based assesment of suitable habitat and patch connectivity for the eurasian lynx. Ecological applications, Vol. 12, No 5, pp. 1469-1483, ISSN 1051-0761

Schadt, S.; Revilla, E.; Wiegand, T.; Knauers, F.; Kaczensky, P.; Breitenmoser, U.; Bufka, L.; O_Cerveny´, J.; Koubek, P.; Huber, T.; Stanisa, C. & Trepl, L. (2002b). Assessing the suitability of central European landscapes for the reintroduction of Eurasian lynx. Journal of Applied Ecology, Vol. 39, pp. 189–203, ISSN 0021-8901

Steyerberg, E.; Eijkemans, M; & Habbema, J. (1999). Stepwise Selection in Small Data Sets: A Simulation Study of Bias in Logistic Regression Analysis. Journal of Clinical Epidemiology, Vol. 52, No. 10, pp. 935–942, ISSN 0895-4356

Swets, J.A. (1988). Measuring the accuracy of diagnostic systems. Science, Vol. 240, 1285–1293, ISSN 0036-8075

Thomas, F. M. & Büttner, G. (1998). Nutrient relations in healthy and damaged stands of mature oaks on clayey soils: two case studies in northwestern Germany. Forest Ecology and Management, Vol. 108, pp. 301-319, ISSN 0378-1127

Tuominen, S. & Pekkarinen, A. (2005). Performance of different spectral and textural aerial photograph features in multi-source forest inventory. Remote Sensing of Environment, Vol. 94, No 2, (January 2005), pp. 256-268, ISSN 0034-4257

Van Staden, V.; Erasmus, B.; Roux, J.; Wingfield, M. & Van Jaarsveld, A. (2004). Modeling the spatial distribution of two important South African plantation forestry pathogens. Forest Ecology and Management, Vol. 187, pp. 61–73, ISSN 0378-1127

Venette, R.C. & Cohen, S.D. (2006). Potential climatic suitability for establishment of *Phytophthora ramorum* within the contiguous United States. Forest Ecology and Management, Vol. 231, pp. 18-26, ISSN 0378-1127

With, K.A. & King, A. (1997). The use and misuse of neutral landscape models in ecology. Oikos, Vol. 79, pp. 219-229, ISSN 0030-1299

With, K.A. (1997). The application of neutral landscape models in conservation biology. Conservation biology, Vol. 11, No 5, pp. 1069-1080, ISSN 0888-8892.

Zhu, Z. & Evans, D. (1994). U.S. forest types and predicted percent forest cover from AVHRR data. Photogrammetric engineering and remote sensing, Vol.60, No 5, pp. 525-531, ISSN 0099-1112

Section 4

Protective and Productive Functions

Soil Compaction – Impact of Harvesters' and Forwarders' Passages on Plant Growth

Roman Gebauer, Jindřich Neruda, Radomír Ulrich and Milena Martinková
Mendel University in Brno, Brno
Czech Republic

1. Introduction

The goal of forestry management is to sustain continual development of forest ecosystems that optimally fulfil their productive and non-productive functions. In order to achieve this goal, the full productive capacity of forest stands needs to be maintained while respecting all the natural processes in the soil, including microbiological organisms, physical properties, nutrient reserves and regeneration processes of the ecosystem.

We need to approach herbs as well as woods holistically, including the root system architecture and functions. Growth of the above-ground system depends on the state of the root system functions, and vice versa. If the conditions for an activity of the root system are limited, the functioning of the above-ground system will be limited too.

During thinning activities in all age groups of forest stands and during the subsequent recovery, progressive harvesting technologies that use mobile means of mechanisation (predominantly harvesters and forwarders) are applied more and more commonly. In contrast to the motomanual technologies that were used in the past, harvesters and forwarders are considerably safer and more productive. However, the passage of heavy machinery on the soil surface causes disruption of the soil environment and mechanical damage to roots. In 1947, it was found that harvesting disrupted soil by modifying its structure and moisture characteristics (Munns, 1947). Despite more than sixty years of research, we still do not fully understand the impact of soil compaction on forest productivity. Due to the global interest in maintaining forest resources and the sustainable development of forest production, a number of conferences have been organised, including the Earth Summit in 1992, which gave rise to the Montreal Process (Burger & Kelting, 1998). At this summit, soil compaction was defined as one of the soil indicators of the forest health state.

Soil compaction is affected by both endogenous and exogenous soil factors. Horn (1988) defined the following endogenous factors as responsible for soil compaction: distribution and size of soil elements, type of clay mineral, type and amount of absorbed cations, content of organic matter, soil structure, soil stabilisation, topsoil material, bulk density of soil, pore continuity and water content. Exogenous factors include the duration, intensity and means of wood harvesting and wood loading. For instance, different machines, or even the same machines with different tyres, differ in their loading and pressure on the soil. Work by Greacen & Sands (1980) and Ole-Meiludie & Njau (1989) support the finding that the compaction rate depends on the concrete soil characteristics, pressure and vibrations of the machines. The rate of soil erosion varies depending on the loading technology and intensity of harvesting. Generally, soil is disrupted by harvest cutting more than it is by selective logging or thinning (Reisinger et al., 1988). The high number of variables leading to soil compaction makes it difficult to find a single parameter that best defines the impact of the passage of a harvester or a forwarder.

2. Harvester and forwarder machinery

Most of the machines currently in use today are heavy and wheeled. The interaction of the wheels with the soil surface in a stand during harvesting and forwarding activities puts pressure on the soil, the intensity of which depends on tyre inflation, toughness and adhesive loading of the traction mechanism. Brais (2001) identified soil compaction by the passage of forestry machines as one of the main factors in soil degradation. Soil compaction during harvesting usually changes the soil structure and moisture conditions by disruption of soil aggregates, decreased porosity, aeration and infiltration capacity, and increased soil bulk density, soil resistance, water interflow, erosion and paludification (Kozlowski, 1999; Grigal, 2000; Holshouser, 2001). Soil compaction may become even more problematic as the weight of harvesters and forwarders increases (Langmaack et al., 2002).

A harvester is a mobile, multi-operational machine that can fell timber, cut branches and chop trunks into assorted lengths in a single cycle (Fig. 1). Individual cut-outs remain in the stand in piles and heaps. The entire process is fully mechanised and automated. Harvesters are classified into four groups based on the kind of undercarriage (wheeled, tracked, walking and combined harvesters). The undercarriage of multi-operational machines has two sections linked by an articulated joint. A forwarder collects the logs made by a harvester and loads them onto a load section of a tractor and forwards them to a storage area (Fig. 2). The main loading function is carried out by a hydraulic crane that reaches 6-10 m with a rotator and a grab.

Fig. 1. Harvester John Deere 1270E with a rotating cab

Fig. 2. Forwarder John Deere 810 D at the platform balance

Forest managers have to concern the total weight of forwarders for particular applications and also the maximal load of tyres has to be observed. Prescribed values for allowable load of tyres according the German Forestry Council (KWF) are given in Table 1. The maximum allowable load of tyres should be up to 4.9 tunes with optimal load up to 4.0 tunes.

Max. weight of forwarder in tunes	Total weight of forwarder (with load) in tunes	Ratio of load on loading part	load of tyre in tunes
8	20	65%	3.2
12	26	65%	4.2
14	30	65%	4.9
16	38	65%	6.2

Table 1. Values for allowable load of forwarder tyres according the German Forestry Council (KWF).

3. Impact of the passage of harvesters and forwarders on soil

The soil compaction that occurs as a consequence of the passage of harvesters and forwarders is connected with significant changes to the soil structure and moisture conditions (Standish et al., 1988; Neruda et al., 2008). Increased bulk density of soil, decreased porosity, decreased water infiltration, increased erosion and changes in plant physiology can all arise from soil compaction. Other changes include the disruption of soil aggregates and loss of pore continuity (Kozlowski, 1999).

3.1 Soil bulk density
Higher soil bulk density is caused by lower porosity and lower water capacity, and it can inhibit root growth (Gebauer & Martinková, 2005). Soil compaction usually occurs in the 30

cm surface layer of soil, which contains the majority of the root biomass (Sands & Bowen, 1978; Kozlowski, 1999) (Fig. 3). The bulk density of soil in the upper layers (0-8 cm) increases by 41-52% after the passage of tractors (Kozlowski, 1999). In the case of a forwarding line, the bulk density of soil in the surface layers (0-10 cm) rose by 15-60% and, in the case of a crossing line, it increased by 25-88% (Lousier, 1990). The compaction decreased in deeper layers; nonetheless, it was recorded even at depths of 30 cm and more. The highest rate of compaction occurred during the first several passages of tractors (Lousier, 1990). The following passages had less effect, but could still lead to rates of compaction that might significantly affect root growth. The critical value of soil bulk density ranges from 1200 to 1400 kg m^{-3}. When this value is exceeded, root growth is reduced in most soil types (Lousier, 1990).

Fig. 3. Superficial root system of a Norway spruce tree showing the majority of the roots growing in the upper soil layer

3.2 Soil porosity

Soil compaction changes the porosity by reducing macroscopic spaces and raising the number of microscopic spaces. The change in porosity affects the balance of soil air and water in pores, which is critical for plant growth. Soil air is a gaseous compound that exists in pores that are not filled with water. Compared with atmospheric air, it includes less oxygen and more CO_2 (ranging from 0.5 – 5% or even higher) (Hillel, 1998). The higher CO_2 content in the soil arises from root respiration and the aerobic decomposition of organic matter. Grable & Siemer (1968) defined the critical value of aeration for plant growth as 10% porosity. Soils with a high content of CO_2 and a low content of oxygen are poorly aerated, and there may even be anaerobic conditions within such soil (Hillel, 1998). A concentration of CO_2 in the soil higher than 0.6 % indicates significant changes to the soil structure that can impact root growth (Güldner, 2002). Our measurements show that this critical value was significantly exceeded in almost all cases after the passage of harvesters and forwarders, and in some cases, the value was exceeded by severalfold (e.g., 1.2% and 3.4% CO_2 in a harvester track as opposed to 0.4 % and 0.5% CO_2 on the surface unaffected by harvesters) (Fig. 4).

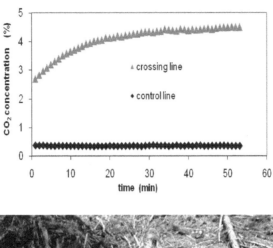

Fig. 4. Upper panel - Concentration of CO_2 in soil air in a crossing line after several harvester passages (soil moisture: 35%). Lower panel – CO_2 measurement in control line with GMP 221 Carbon dioxide probe (Vaisala, Finland)

3.3 Water infiltration and erosion

Soil compaction is often related to the creation of crust, causing decreased water infiltration and ultimately increasing water runoff (Malmer & Grip, 1990). In the places where water runoff is not possible (e.g., holes after passage, terrain depressions), there is weak drainage, which causes local inundation (Jim, 1993) (Fig. 5). Experiments have shown that harvesters and forwarders can accelerate the rate of surface erosion from 2 to 15 times, compared with unpassaged soil and 85% of the total surface erosion appears in the first year after disruption (Lousier, 1990).

We should consider the soil capability i.e. the ability of soil to cope with external forces, which can cause permanent or temporal deformation, when heavy machines are moving in the forest. The rut depth from 15 - 50 cm (according the soil humidity) brings high ecological risk (Fig. 6). The soil capability of different soil types is given in Table 2.

Fig. 5. A case of unsuitable preparation of a site with disruption of soil aggregates. If an Eco-Baltic wheeled track had been used, the lines would not have been cut to a depth of 50 cm and deeper along the way.

Fig. 6. A case of rut depth up to 25 cm, which is a point when an ecological risk may appear.

3.4 Plant physiology
3.4.1 Disorders in photosynthesis and water regime

Heavy compaction leads to a variety of physiological disorders in plants. Roots react to soil compaction by increasing demand for photosynthates (Zaerr & Lavender, 1974), which are needed to support the metabolism required to overcome the increased soil resistance to elongation growth. The physiological cost of recovering the functions of fine roots may be as high as 70% of the accessible carbon flow (Ågren et al., 1980; Vogt et al., 1996). Kozlowski (1999) found that the increased carbon flow due to soil compaction leads to an overall decrease in photosynthesis. This is a result of reduced foliage surface, which is an outcome of reduced water intake caused by changes in the soil structure and moisture conditions (Arvidsson & Jokela, 1995). Therefore, a plant might not have enough energy to reconstruct its root system, and the growth of roots as well as the above-ground parts stagnate or even die. Reduced foliage surface is a reaction to a water deficit in the leaves, which is brought about by soil compaction and may lead to the closing of pores and further loss of photosynthesis (Masle & Passioura, 1987).

degree of resistance	soil capability	rut depth, soil consistence	soil taxonomy
1	extremely low dry: 30 -50 kPa wet: 5-12 kPa	≥ 35 cm, incohesive, strongly crumble, slush	Histosols, Gleysols
2	very low dry: 50-140 kPa wet: 12-22 kPa	25-35 cm crumbly, clay, loam, very soft	Stagnosols, gleyic Stagnosols
3	reduce dry: 140 - 300 kPa wet: 18-50 kPa	15-25 cm hardly dig, loam, sandy clay, soft	Cambisols, Luvisols, Fluvisols - subtype - gleyic
4	slightly reduce dry: 300-600 kPa wet: 50-80 kPa	7-15 cm hardly dig, solid, sandy loam	dry and slightly wet Cambisols, Luvisols, Regosols, Chernozems
5	bearable dry: > 600 kPa wet: 80-120 kPa	< 7 cm solid, hard, stony	Podzols, Leptosols

Table 2. Soil capability measured as a rut depth after one passage of the special forest tractor (LKT 80) with inflation of tyres 200 kPa. Dry and wet means humidity of sandy and loam-sandy soil 4-8 % and 18-30%; sandy-loam and loam soil 8-15% and 35-45%; clay-loam and clay soil 15-25% and 45-55%, respectively.

3.4.2 Disorders in nutrient uptake
Often, extreme soil compaction leads to reduced absorption of mineral nutrients by the roots, especially nitrogen, phosphorus and potassium. Nutrient uptake is reduced as a result of the loss of minerals from soil, reduction of root access to nutrients and decreased root capacity for nutrient intake (Kang & Lal, 1981; Kozlowski & Pallardy, 1997). A reduction of nutrient uptake caused by soil compaction in the upper as well as deeper soil layers (Kozlowski, 1999) might be the reason for different reactions to the compaction among species, as some have higher nutrient demands than others.

3.4.3 Effects on mycorrhizas and plant hormones
Soil compaction also affects the structure, development and function of mycorrhizas (Entry et al., 2002) and causes changes in the levels of stress hormones in plants, mainly abscisic acid and ethylene (Kozlowski, 1999).

3.4.4 Respiration disorders
Soil compaction induces hypoxia, which is related to the reduction of aerobic micro-organism activity and an increase of denitrification. As compaction increases, reduction of macro-pores enhances the development of anaerobic spaces (Torbert & Wood, 1992). Insufficient aeration of compacted soils leads to anaerobic respiration in roots and insufficient energy for maintaining the basic root functions, namely nutrient uptake (Kozlowski & Pallardy, 1997).

4. Impact of compaction on plant growth

Several studies have shown that tree growth and wood production decrease with increasing compaction (Froehlich, 1976; Cochran & Brock, 1985). Growth inhibition as well as the death of woody plants caused by soil compaction has been documented in zones of recreation, harvesting areas (Sand & Bowen, 1978; Cochran & Brock, 1985), agro forestry (Wairiu et al., 1993) and tree nurseries (Boyer & South, 1988).

Soil compaction strongly reduces plant growth as it limits root growth (Rosolem et al., 2002; Gebauer & Martinková, 2005). There is a non-linear relationship between root elongation and soil resistance in the majority of plants (Misra & Gibbons, 1996). Because compaction usually occurs in the upper soil levels, species with a surface root system are disadvantaged (Godefroid & Koedam, 2003). Generally in the case of large trees, root growth is limited by increasing soil bulk density and excessive soil resistance (typical in dry and skeletal soils) or insufficient aeration if the soil is heavily saturated by water (Greacen & Sands, 1980). The greater the root growth reduction and the smaller the soil space occupied by roots, the slower the growth of a tree in its above-ground parts (Halverson & Zisa, 1982; Tuttle et al., 1988).

The exposure of roots to mechanical pressure induces a number of physiological changes that have been well described on the macroscopic level. For example, the elongation growth decreases, and the response period varies from several minutes (Sarquis et al., 1991; Bengough & MacKenzie, 1994) to many hours (Eavis, 1967; Croser et al., 1999). The root tip generally rounds, becoming concave, the root width behind the meristem increases and the root meristem and the elongation zone shorten (Eavis, 1967; Croser et al., 2000). The data on root thickening behind the root tip demonstrate the effects of long-term mechanical pressure on the root tips (Abdalla et al., 1969; Martinková & Gebauer, 2005). The growth of roots is reported to be a more sensitive indicator of soil disruption than the growth of the above-ground parts (Singer, 1981; Heilman, 1981) because the reduction of root growth precedes the phase when the extreme soil resistance is achieved (Eavis, 1967; Russell, 1977; Simons & Pope, 1987).

The critical value of soil resistance that can lead to significant physiological changes is measured by penetrometers (Atwell, 1993; Greacen & Sands, 1980) (Fig. 7), which better express conditions of root growth as penetrometers also measure the influence of bulk density and soil moisture (Siegel-Issem, 2002). Heavy, humid soils are more easily penetrated by roots due to lower soil resistance, while in arid soils of the same density, the growing resistance limits root growth. Critical values of compaction, expressed by penetrometric soil resistance, for different kinds of soil are listed in Table 3. It has been determined that a soil resistance of 2.0 MPa or more causes root shortening in most plant species (Atwell, 1993). The critical soil resistance on compacted sands limiting root growth measured for *Pinus radiata* was 3.0 MPa (Sands et al., 1979). However, roots usually have a lower resistance to soil penetration than the resistance measured by penetrometers, due to the radial expansion and smaller diameter of roots and the ability to curl and minimise friction by means of polysaccharide slime.

Only a few studies, mainly using herbs, have measured the soil resistance against roots directly in soil (Eavis, 1967; Misra et al., 1986; Bengough & Mullins, 1991; Clark & Barraclough, 1999). Roots were found to be capable of exert the outer pressure from 0.9 to 1.3 MPa (Gill & Miller, 1956; Barley, 1962; Taylor & Ratliff, 1969). Eavis (1967) demonstrated that elongation of roots in peas was reduced by 50% at a pressure of 0.3 MPa. Our measurements show that soil compaction causes reduced root elongation growth in Norway spruce by 50% compared with control seedlings (Gebauer & Martinková, 2005) (Fig. 8). In the case of one-year-old buds of Scotch Pine (*Pinus sylvestris*), the soil compaction did not have a significant impact, but for Macedonian Pine (*Pinus peuce*) of the same age, the root growth was negatively affected by soil compaction (Mickovski & Ennos, 2002; 2003). The authors of this study reasoned that the weak impact on *Pinus sylvestris* was due to the fact that its roots have thinner diameters than those of *Pinus peuce*.

Fig. 7. Measurement of soil resistance by penetrometer

Soil type	penetrometric soil resistance (MPa)
Sandy loam and sand	more than 4
Sandy clay	4 – 3.7
Silt	3.7 – 3.5
Silty clay	3.5 – 3.2
Clay	less than 3.2

Table 3. Critical values of penetrometric resistance of soil types

The above study shows that compaction significantly reduces plant growth; yet, other studies show that the compaction of soils with a coarse structure (sandy soils) might have a positive impact on the growth of conifers. This contradiction may be because the compaction of sandy soils creates microscopic spaces and enhances water retention in the soil (Troncoso, 1997; Gomez et al., 2002; Siegel-Issem, 2002). Mild soil compaction in sand supports the contact between roots and soil, resulting in higher absorption of water and nutrients (Gomez et al., 2002; Alameda & Villar, 2009). Alameda & Villar (2009) found that a mild compaction positively affected the growth of 53% of seedlings from 17 species (including both foliage and coniferous seedlings) growing in controlled conditions. Miller et al. (1996) found that in forwarding lines with an increased soil bulk density of 40% or more, growth was not affected at all, and 8-year-old seedlings of *Pseudotsuga menziessi* and *Picea sitchensis* survived.

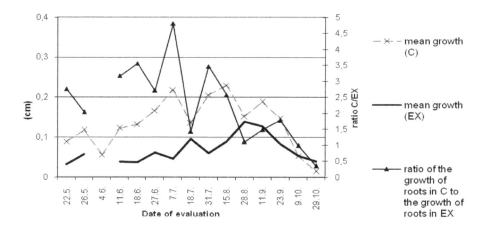

Fig. 8. Root growth and dynamics of Norway spruce seedlings grown in control non-compacted root boxes (C) and in root boxes exposed to a long-term pressure of 5.1 kPa (EX). A C/EX ratio above one indicates higher root growth in the non-compacted soil (Gebauer & Martinková, 2005).

In general, soil compaction is a stress factor that negatively affects the growth of plants, but the rates of compaction and differences among soil types need to be taken into account in these analyses (Kozlowski, 1999; Alameda & Villar, 2009). For instance, Alameda & Villar (2009) showed that growth increases in most seedlings grown in a sandy substrate with rising compaction of 0.2-0.6 MPa, but exceeding this value generally led to a reduction in growth.

5. Recording of harvesters' and forwarders' pressures on soil

During the passage of heavy vehicles on unsurfaced soil, the soil environment gets disrupted and roots are mechanically injured. A method for measuring and recording the immediate pressure on soil was developed and tested by the institute of Forest and Forest Products Technology of MENDELU in Brno (Czech Republic). This method is applicable in forest stands that grow on mild soil surfaces where large and extremely heavy machines (forwarders) pass. Pressure sensors were placed in the soil near the surface, and a unique measuring chain was used to measure the immediate pressure on the soil.

The pressure on a concrete point (e.g., a root or stress sensor) exerted by a wheel is short-lived (approx. 0.04 s) and has a stress impulse character (Fig. 9). The impulse does not have a permanent value, so its rise, apex and fall can be clearly observed. The apex values of stress impulses were used in measuring the stress on the soil. This method is helpful for determining suitable precautions in forestry management, e.g., the effect of different covers on soil protection and the optimal height of the layer. Moreover, this method establishes the optimal inflation of tyres because over-inflated tyres, even the low-pressure type, lead to higher stress on the soil.

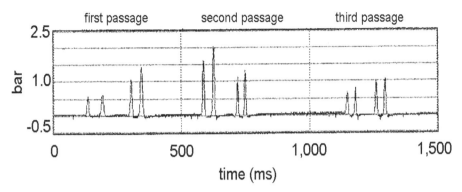

Fig. 9. Recording of the measurement of soil pressures during a forwarder's passage

6. Recovery of compacted soil

Revitalisation and amelioration of compacted soil is a long-term process and it is not known if it is fully achievable (Heninger et al., 2002). The regeneration period after the compaction may be less than 10 years near the soil surface (Thorud & Frissel, 1976; Lowery & Schuler, 1994), but others claim it could last several decades (Wert & Thomas, 1981; Jakobsen, 1983; Froehlich et al., 1985). It is necessary to fully understand the process of compaction, its impact on soil and plant growth and to find means and technologies that minimise the influence of compaction (if at all possible).

The recovery of compacted soil is a result of the combination of root activity, freeze-melt cycles and humid-dry cycles (Reisinger et al., 1988). After a period of 5 years, the bulk density of the surface, which consists of fine sandy-silt soil, was higher by 12% in former lines compared with places unaffected by the compaction (Lockaby & Vidrine, 1984).

The revitalisation of compacted soil also depends on the content of the organic matter in the soil, as it has a significant impact on the soil structure, aeration, water retention and chemical properties. Soil bulk density and porosity increase or decrease with the growing content of organic matter (Childs et al., 1989). Differences of 2-5% may significantly affect soil properties such as bulk density and porosity in sandy soils (Rawls, 1983).

We do not know of any ways to revitalise compacted forest soil on a large scale by technical means or technologies. Thus, it is necessary to prevent soil compaction by forestry management.

7. Prevention of soil compaction

The rate of soil compaction varies considerably depending on the method of felling, the type of soil preparation, the terrain conditions, the timing of the activity and the preparation and personal responsibility of the workers. Soil disruption by harvesting is also affected by soil conditions during the activity (e.g., soil resistance, humidity, frost, snow cover), concrete features of the activity (e.g., frequency of passages) and the impact (stress and vibration) on the soil by harvesters and forwarders.

During the movement of heavy tractors through areas with little bearing capacity of the subsoil, permanent deformations of terrain (lines 20 – 50 cm deep) arise. Even though these lines might be relatively short (5 - 15 m), they make the given section permanently impassable and inaccessible to wheeled or tracked tractors. Such sections include friable sand, drift sand, wet sand, permanently flooded places, passages to bridge inundated areas of watercourses, ford beds, passages in marshy or peaty terrain and dumps. Subsoils at extreme risk include clay soils, because they absorb high amounts of water and their bearing capacity is problematic in the spring and autumn. This highlights the necessity of clearing such a stand prior to activities on weakly bearing terrain.

Preparation of weakly bearing surfaces for harvesting is carried out in two ways:

1. The forwarding route is reinforced with additional material.
2. The road structure is temporarily reinforced (gabions, plastic mobile grids, plastic mobile boards, low-pressure tyres, route reinforcing –old used forest fences, harvesting waste). The extent of the reinforcement needed mainly depends on the axle pressure of the vehicle, construction and strength of the road, mechanical and physical properties of the terrain and the required number of passages of the vehicle.

The advantage of grids and screens is that they are quick and easy to use (Fig. 10). Local reinforcement of a road by means of screens can be achieved along the whole route for minimal costs. After pressing through the bottom layers of the soil, the skid of the wheels on the screen falls rapidly too. The producer recommends 8 tons as the maximal bearing capacity of screens; however, they have been successfully tested with forwarders loaded with 10 - 15 tons (Schlaghamersky, 1991; Ulrich & Schlaghamersky, 1997). Placement of a screen can open the way to a very wet biotope without soil damage by deep lines. One disadvantage of screens is that they cannot be placed directly on unprepared terrain; the lines resulting from the wheels need to be filled with brushwood or harvesting waste, for example. After a long period, soil gets through the screen and needs to be removed by a blade.

Fig. 10. Plastic mobile grids are quick and easy to use.

Besides the proper preparation of the terrain for the passage, there are other ways of minimising soil compaction by the modification of harvesting technologies. For example, the application of lighter technology (Jansson & Wästerlund, 1999), lower inflation of tyres (Canillas & Salokhe, 2001), placement of harvesting waste in locations where harvesting and forwarding is planned (Hutchings et al., 2002) (Fig. 11), harvesting in winter on frozen soil (Alban et al., 1994), planting species tolerant to compaction (Bowen, 1981; Ruark et al. 1982) and limitation of drawing logs using a winch can all help reduce soil compaction. Limitation of the number of passages would not help because 80% of soil compaction occurs during the first passage (Holshouser, 2001). The most efficient precaution is prevention against soil compaction, as the other methods might be ineffective and, furthermore, could do harm to the roots (Howard et al., 1981).

Fig. 11. Placement of harvesting waste in places of forwarders' and harvesters' passages is one way to minimise soil compaction.

8. Conclusion

The passage of forestry machines causes soil compaction, leading to significant changes in the soil structure and moisture conditions. When soil is compacted, soil bulk density increases, porosity and water infiltration decrease, erosion speeds up, and all of these processes lead to changes in plant physiology. Photosynthesis, transpiration, nutrient uptake, mycorrhizas and plant hormones are all possible avenues for these changes.

Soil compaction is influenced by endogenous soil factors (distribution and size of soil elements, soil bulk density, pore continuity, water content, etc.) as well as exogenous factors (choice of equipment, loading of wood, length of loading, intensity and means of harvesting, site preparation, etc.). When soil is compacted, the soil resistance grows; resistance over 2.0 MPa, as measured by penetrometer, limits elongation root growth in most plant species. Our measurements have shown that this critical value is often exceeded when forestry machines pass through an area without any preparation of the site.

Poor aeration of soil caused by soil compaction also prevents the development of root systems and limits the water penetrability of roots. Our measurements show that the critical value of CO_2 in the soil air (defining the rate of aeration) was exceeded as a result of the passage of forestry machines in almost all cases. To establish the optimal inflation of tyres the pressure sensor (a sensor developed and tested by us) was found to be very useful tool.

This sensors are also applicable in forestry management because it aids in the determination of suitable precautions, e.g., whether the soil surface is covered with a sufficient layer of brushwood.

Although compaction is usually considered to be a factor of growth deceleration, some studies of conifers show that compaction of certain soils with a coarse structure (sandy soils) may, on the contrary, enhance growth due to the multiplication of microscopic pores, thus increasing the soil's capability to retain a higher amount of water.

Since the revitalisation and amelioration of compacted soil is a long-term process, and it is not unknown if it is fully achievable, compaction should be minimised as much as possible. Its minimisation could be achieved by the modification of technologies in forestry activities; for instance, by using lighter machines, reducing tyre pressure, placing harvesting waste in places where forestry machines are expected to pass, harvesting in the winter on frozen soil and controlling tractor movement. We should also mention that human factors play often a critical role in the soil compaction.

9. Acknowledgement

The authors mainly thanks to the support of partial research projects No. MSM 6215648902 "Rules of the management and optimisation of species structure of forests in antropically changing conditions of hilly areas and highlands" and "Risks of a decline of spruce stands in highlands and hilly areas" and internal grant projects no. 19/2010 "Harvester forwarder systems and power engineering – harmonisation of forestry activities with the environment" and no. 12/2010 "Using of genetic information in forestry botanics, woody plant physiology, dendrology and geobiocenology".

10. References

Abdalla, A.M.; Hettiaratchi, D.R. & Reece, A.R. (1969). The mechanics of root growth in granular media. *Agric Eng Res* 14:236-248.

Alameda, D. & Villar, R. (2009). Moderate soil compaction: implications on growth and architecture of 17 woody plant seedlings. *Soil Till Res* 103:325–331.

Alban, D.H.; Host, G.E.; Elioff, J.D. & Shadis D. (1994). Soil and vegetation response to soil compaction and forest floor removal after aspen harvesting. *US Department of Agriculture, Forest Service, Res. Pap. NC-315*. North Central Forest Experiment Station, St. Paul, MN. 8 pp.

Arvidsson, J. & Jokela, W.E. (1995). A lysimeter study of soil compaction on wheat during early tillering. II. Concentration of cell constituents. *New Phytol* 115:37-41.

Atwell, B.J. (1993). Response of roots to mechanical impedance. *Environ Exp Bot* 33:27-40.

Ågren, G.; Axelsson, B.; Flower-Ellis, J.G.K.; Linder, S.; Persson, H.; Staaf, H. & Troeng, E. (1980). Annual carbon budget for young Scots pine. – In: *Structure and Function of Northern Coniferous Forest,* T. Persson, (Ed.), An Ecosystem Study -Ecol Bull (Stockholm) 32:307-313.

Barley, K.P. (1962). The effects of mechanical stress on the growth of roots. *J Exp Bot* 13: 95-110.

Bengough, A.G. & Mullins, C.E. (1991). Penetrometer resistence, root penetration resistence and root elongation rate in two sandy loam soils. *Plant Soil* 132:59-66.

Bengough, A.G. & MacKenzii, C.J. (1994). Simultaneous measurement of root force and elongation rate for seedling pea roots. *J Exp Bot* 45:95-102.

Bowen, H.D. (1981). Alleviating mechanical impedance. In: *Modifying the Root Environment to Reduce Crop Stress,* G.F. Arkin & H.M. Taylor (Eds.). St. Joseph, MI: ASAE. 21-57 pp.

Boyer, J.N. & South D.B. (1988). Date of sowing and emergence timing affect growth and development of loblolly pine seedlings. *New For* 2:231-246.

Brais, S. (2001). Persistence of soil compaction and effects on seedlings growth in northwestern Quebec. *Soil Science Society of America Journal* 65:1263-1271.

Burger, J. A. & Kelting, D. L. (1998). Soil quality monitoring for assessing sustainable forest management. In: *The contribution of soil science to the development of and implementation of criteria and indicators of sustainable forest management.* E. A, Davidson; M. B. Adams & K. Ramakrishna (Eds). SSSA Special Publication Number 53. Madison, WI: Soil Science Society of America. 17-52 pp.

Canillas, E.C. & Salokhe W.M. (2001). Regression analysis of some factors influencing soil compaction. *Soil and Tillage Research* 61: 167-178.

Childs, S.W.; Shade, S.P; Miles, D.W.; Shepard, E. & Froehlich, H.A. (1989). Management of soil physical properties limiting forest productivity. In: *Maintaining the long-term productivity of Pacific Northwest forest ecosystem,* D.A. Perry et al. (Eds.). Timber Press, Portland, OR, USA.

Clark, L.J. & Barraclough, P.B. (1999). Do dicotyledons generate greater maximum axial root growth pressures than monocotyledons? *J Exp Bot* 50:1263-1226.

Cochrad, P.H. & Brock, T. (1985). Soil compaction and initial height growth of planted ponderosa pine. *USDA For Serv Res Note PNW-434.*

Croser, C.; Bengough, A.G. & Pritchard, J. (1999). The effect of mechanical impedance on root growth in pea (*Pisum sativum*). I. Rates of cell flux, mitosis, and strain during recovery. *Physiol Plant* 107:277-286.

Croser, C.; Bengough, A.G. & Pritchard, J. (2000). The effect of mechanical impedance on root growth in pea (*Pisum sativum*). II. Cell expansion and wall rheology during recovery. *Physiol Plant* 109:150-159.

Eavis, B.W. (1967). Mechanical impedance to root growth. *Agricultural Engineering Symposium.* Silsoe. Paper 4/F/39:1-11.

Entry, J.A.; Rygiewicz, P.T.; Watrud, L.S. & Donnelly, P.K. (2002). Influence of adverse soil conditions on the formation and function of arbuscular mycorrhizas. *Advences in Environmental Research* 7:123-138.

Froehlich, H.A. (1976). The effect of soil compaction by logging on forest productivity, part 1. *USDI Bureau of Land Management.* Final Report, Contract No. 53500-CT4-5(N).

Froehlich, H.A.; Miles, D.W.R. & Robbins, R.W. (1985). Soil bulk density, recovery on compacted skid trails in central idaho. *Soil Sci Soc Am J* 4:1015–1017.

Gebauer, R. & Martinková, M. (2005): Effects of pressure on the root systems of Norway spruce plants (*Picea abies* [L.] Karst.). *Journal of Forest Science* 51:268-275.

Gill, W.R. & Miller, R.D. (1956). A method for study of the influence of mechanical impedance and aeration on seedling roots. *Soil Science of America Proceedings* 20:154-157.

Godefroid, S. & Koedam N. (2003). How important are large vs. Small forest remnants for the conservation of the woodland flora in an urban kontext? *Global Ecology and Biogeography* 12:287-298.

Gomez, G.A.; Powers, R.F.; Singer, M.J. & Horwath, W.R. (2002). Soil compaction effects on growth of young ponderosa pine following litter removal in California's Sierra Nevada. *Soil Sci Soc Am J* 66:1334-1343.

Grable, A.R. & Siemer, E.G. (1968). Effects of bulk density, aggregate size, and soil water suction on oxygen diffusion, redox potentials, and elogation of corn roots. *Soil Sci Soc Am Proc* 32:180-186.

Greacen, E.L. & Sands, R. (1980). Compaction of Forest Soils: A Review. *Aust J Soil Res* 18:163-189.

Grigal, D.F. (2000). Effects of extensit forest management on soil produktivity. *Forest Ecology and Management* 138:167-185.

Güldner, O. (2002). Untersuchungen zu Bodenveränderungen durch die Holzernte in Sachsen und Entwicklung eines Konzepts zur ökologisch verträglichen Feinerschliessung von Waldbeständen. *Tagungsbericht der Sektion Forsttechnik des Verbandes Deutscher Forstlicher Versuchsanstalten*, Sopron, Ungarn, 10 p.

Halverson, H.G. & Zisa, R.P. (1982). Measuring the response of conifer seedlings to soil compaction stress. *USDA For Serv Res Pap NE-509*.

Heilman, P. (1981). Root penetration of Douglas fir seedlings into compacted soil. *For Sci* 27:660-666.

Heninger, R.; Scott, W.; Dobkowski, A.; Miller, R.; Anderson, H. & Duke, S. (2002). Soil disturbance and 10-year growth response of coast Douglas-fir on nontilled and tilled skid trails in the Oregon Cascades. *Can J For Res* 32:233–246.

Hillel, D. (1998). *Environmental Soil Physics*. Academia Press. USA.

Holshouser, D.L. (Ed.). (2001). *Soybean Production Guide*. Tidewater Agricultural Research and Extension Center, Information Series No. 408.

Horn, R. (1988). Compressibility of arable land. In: *Impact of water and external forces on soil structure*, J. Drescher et al. (Eds.) Catena suppl. 11, Catena Verlag, Germany. p.53-71.

Howard, R.F.; Singer, M.J. & Frantz G.A. (1981). Effects of soil properties, water content, and compactive effort on the compaction of selected California forest and range soils. *Soil Sci Soc AmJ* 45:231-236.

Hutchings, T.R.; Moffat, A.J. & French, C.J. (2002). Soil compaction under timber harvesting machinery: A preliminary report on the role of brash mats in its prevention. *Soil Use and Management* 18:34–38.

Jakobsen, B.F. (1983). Persistence of compaction effects in a forest Kraznozen. *Aust J For Res* 13: 305-308.

Jansson, K.J. &Wästerlund, I. (1999). Effect of tradic by lightwright forest machinery on the growth of young Picea abies trees. *Scandinavian Journal of Forest Research* 14: 581-588.

Jim, C.Y. (1993). Soil compaction as a constraint to tree growth in tropical and subtropical urban habitats. *Environ Conserv* 20:35–49.

Kang, B.T. & Lal R. (1981) Nutrient losses in water runoff from a tropical watershed, In: Tropical agricultural hydrology, R.Lal & E. W.Russell (Eds.). Wiley, Chichester. *pp.* 119–130.

Kozlowski, T.T. (1999). Soil compaction and growth of woody plants. *Scandinavian Journal of Forest Research* 14:596-619.

Kozlowski, T.T. & Pallardy, S.G. (1997). Physiology of Woody Plants, 2nd edition. Academic Press, San Diego.

Langmaack, M.; Schrader S.; Rapp-Bernhardt, U. & Kotzke K. (2002). Soil structure rehabilitation of arable soil degraded by compaction. *Geoderma* 105:141-152.

Lockaby, B.G. & Vidrine, C.G. (1984). Effect of logging equipment traffic on soil density and growth and survival of young loblolly pine. *South J App. For* 8:109-112.

Lousier, J.D. (1990). Impacts of Forest Harvesting and Regeneration on Forest Sites. *Land Management*. Report Number 67.

Lowery, B. & Schuler, R.T. (1994). Duration and effects of compaction on soil and plant growth in Wisconsin. *Soil Till Res* 29:205-210.

Malmer, A. & Grip, H. (1990). Soil disturbance and loss of infiltrability caused by mechanized and manual extraction of tropical rainforest in Sabah, Malaysia. *Forest Ecology and Management* 38:1–12.

Martinková, M. & Gebauer, R. (2005). Problematika mechanického poškození kořenů rostlin. In: *Metody pro zlepšení determinace poškození kořenů stromů ve smrkových porostech vyvážecími traktory. I. Výběr a ověření metod (In Czech)*. J. Neruda (Ed.): Monograph. Brno. p. 7-10.

Masle, J. & Passioura, J.B. (1987). The effect of soil strength on the growth of young wheat plants. *Australian Journal of Plant Physiology* 14:643–656.

Mickovski, S.B. & Ennos, A.R. (2002). A morphological and mechanical study of the root systems of suppressed crown Scots pine *Pinus silvestris*. *Trees Struct Funct* 16:274-280.

Mickovski, S.B. & Ennos A.R. (2003). Anchorage and asymmetry in the root system of *Pinus peuce*. *Silva Fenn* 37:161-173.

Miller, R.E.; Scott W. & Hazard, J.W. (1996). Soil compaction and conifer growth after tractor yarding at three coastal Washington locations. *Can J For Res* 26:225–236.

Misra, R.K.; Dexter, A.R & Alston, A.M. (1986). Maximum axial and radial growth pressures of plant roots. *Plant Soil* 95:315-326.

Misra, R.K. & Gibbons, A.K. (1996). Growth and morphology of eucalypt seedling-roots, in relation to soil strength arising from compaction. *Plant Soil* 182:1–11.

Munns, E.N. (1947). Logging can damage soil. *J For* 45:513.

Neruda, J.; Čermák, J.; Naděždina, N.; Ulrich, R.; Gebauer, R.; Vavříček, D.; Martinková, M.; Knott, R.; Prax, A.; Pokorný, E.; Aubrecht, L.; Staněk, Z.; Koller, J. & Hruška J. (2008). *Determination of damage to soil and root systems of forest trees by the operation of logging machines*. Mendel University in Brno, Brně. 138 p.

Ole-Meiludiem R.E.L. & Njau, W.L.M. (1989). Impact of logging equipment on water infiltration capacity at Olmotonyi, Tanzania. *For Ecol Manage* 26:207–213.

Rawls, W.J. (1983). Estimating soil bulk density from particle size analysis and organic matter content. *Soil Sci* 135:123-125.

Reisinger, T.W.; Simmons, G.L. & Pope P.E. (1988). The impact of timber harvesting on soil properties and seedling growth in the south. *S J App For* 12:58–67.

Rosolem, C.A.; Foloni, J.S.S. & Tiritan, C.S. (2002). Root growth and nutrient accumulation in cover crops as affected by soil compaction. *Soil and Tillage Research* 65:109-115.

Ruark, G.A.; Mader, D.L. & Tatter, T.A. (1982). A composite sampling technique to assess urban soils under roadside trees. *J Arboric* 8:96-99.

Russell, R.S. (1977). Mechanical impedance of root growth. In: Plant Root Systems: Their function and interaction with soil. R.S. Russell (Ed.). McGraw-Hill Book Company, UK. 98-111 pp.

Sands, R. & Bowen, G.D. (1978). Compaction of sandy soils in Radiata pine forests. II. Effects of compaction on root configuration and growth of radiata pine seedlings. *Aust J For Res* 8:163-170.

Sands, R.; Greacen, E.L. & Gerard, C.J. (1979). Compaction of sandy soils in Radiata pine forests. I. A penetrometer study. *Aust J Soil Res* 17:101-113.

Sarquis, J.I.; Jordan, W.R. & Morgan, P.W. (1991). Ethylene evolution from maize (*Zea mays* L.) seedling roots sholte in response to mechanical impedance. *Plant Physiol* 96:1171-1177.

Schlaghamersky,A. (1991). Entwicklung von Methoden und einfachen Geraeten zur Bestimmung der Oberflaechentragfaehigkeit von Waldboeden. *Bericht Fachhochschule Hildesheim-Holzminden.* pp.74

Siegel-Issem, C.M. (2002). Forest productivity as a function of root growth opportunity. Thesis submitted to the faculty of the Virginia Polytechnic Institute and State University. Blacksburg, Virginia. 86 p.

Simmons, G.L. & Pope, P.E. (1987). Influence of soil compaction and vesicular-arbuscular mycorrhizae on root growth of yellow poplar and sweetgum seedlings. *Can J For Res* 17:970-975.

Singer, M.J. (1981). Soil compaction – seedling growth study. *Final report to USDA Forest Service, Pacific Southwest Region.* Coop. Agree. USDA-7USC-2202. Suppl. 43.

Standish, J.T.; Commandeur, P.R. & Smith, R.B. (1988). Impacts of forest harvesting on physical properties of soils with reference to increased biomass recovery — a review. *Inf Rep BC-X-301, B.C.* Canadian Forest Service Pacific Forestry Research Centre.

Taylor, H.M. & Ratliff, L.F. (1969). Root elongation rates of cotton and peanuts as a function of soil strength and soil water content. *Soil Sci* 108:113-119.

Thorud, D.B. & Frissel, S.S Jr. (1976). Time changes in soil density following compaction under an oak forest. *Minnesota Forestry Notes No. 257.* Univ. of Minnesota, St. Paul, MN.

Torbert, H.A. & Wood, C.W. (1992). Effects of soil compaction and water-filled pore space on soil microbial activity and N losses. Commun. *Soil Sci Plant Anal* 23:1321–1331.

Troncoso, G. (1997). Effect of soil compaction and organic residues on spring-summer soil moisture and temperature regimes in the Sierra National Forest. *M.S. thesis.* Oregon State Univ., Corvallis, OR.

Tuttle, C.L.; Golden, M.S. & Meldahl, R.S. (1988). Soil compaction effects on *Pinus taeda* establishment from seed and early growth. *Can J For Res* 18:628-632.

Ulrich,R., Schlaghamersky,A.(1997): Tool for bearability increase of the forest tracks (In Czech). Usable model. *Úřad průmyslového vlastnictví ČR.* Číslo zápisu:5479, MPT: E01 C9/02

Vogt, K.A.; Vogt, D.J.; Palmiotto, P.A.; Boon, P.; O´hara, J. & Asbjotnsen, H. (1996). Review of root dynamics in forest ecosystems grouped by climate, climatic forest type and species. *Plant Soil* 187:159-219.

Wairiu, M.; Mullins, C.E. & Campbell, D. (1993). Soil physical factors affecting the growth of sycamore (*Acer pseudoplatanus* L.) in a silvopastoral system on a stony upland soil in North-East Scotland. *Agrofor System* 24:295–306

Wert, S. & Thomas, B.R. (1981). Effects of skid roads on diameter, height, and volume growth in Douglas-fir. *Soil Sci Soc Am J* 45:629-632.

Zaerr, J.B. & Lavender, D.P. (1974). The effects of certain cultural and environmental treatments upon growth of roots of Douglas fir (*Pseudotsuga menziessii* /Mirb. / Franco) seedlings. *Proceedings of International Symposium on Ecology and Root Growth.* Biol. Gessellschaft. Postdam, Germany, pp. 27 – 32.

10

Ecological Consequences of Increased Biomass Removal for Bioenergy from Boreal Forests

Nicholas Clarke
Norwegian Forest and Landscape Institute
Norway

1. Introduction

The increased use of renewable energy sources, including forest biomass, in energy consumption is a marked characteristic in many countries' current energy policies. Use of forest biomass for energy is supported as a sustainable form of energy that contributes to social welfare, local development and forest economy. Thus, in Europe there is a sharp increase in demand for wood as a source of renewable energy as well as for production of wood products. Forest inventories show that standing stock as well as annual growth would allow an increased use of the existing forest resource.

In conventional stem-only timber harvesting (SOH), where branches and tops are left in the forests, the organic material will decay on the site and nutrients are thus returned to the biogeochemical cycle. In whole-tree harvesting (WTH), branches and tops are removed, although in practice the amount removed is about 60-80% (Helmisaari et al., 2011). As a large part of the nutrients in trees are located in the foliage and branches, removing these will reduce the supply of nutrients and organic matter to the soil. In the longer term, this might increase the risk for future nutrient imbalances and reduced forest production (Egnell & Leijon 1999; Raulund-Rasmussen et al., 2008; Worrell & Hampson, 1997), as well as changes in species composition and biodiversity (Jonsell, 2007). In some countries, such as Finland, stumps may also be harvested, although this will not be considered here.

Forests provide a number of environmental services, such as water protection, carbon sequestration and biological diversity, which need to be maintained both during and after harvesting. Removal of forest residues after harvesting could increase the risk for adverse effects on these services. Thus, there is a potential for conflict between such goals as increased use of forest resources for bioenergy and rural employment on the one hand, and protection of ecosystem services together with long-term site sustainability on the other. In order to minimise the potential for conflict, legislation, certification systems and management guidelines have been developed. However, for these to be effective, there has to be a scientific basis, and there is at present insufficient knowledge about which factors determine the contrasting effects found in field experiments on increased biomass removal (see below), or of how variation in these controlling factors affects long-term site sustainability. This review will address the current state of knowledge regarding sustainable removal of branches and tops for bioenergy from boreal forest ecosystems.

2. Effects of harvesting intensity on soil and water

Nutrient depletion is the major environmental concern regarding WTH as compared with SOH, as this is relevant not only environmentally but also economically due to the risk for reduced growth in the next rotation. As stated above, a large portion of tree nutrients are in the foliage and branches, so removing these from the forest will also remove the nutrients. If these nutrients are not replaced, either by weathering, deposition or fertilisation, reduced growth in the next rotation may result. This risk will vary greatly, depending on site nutrient status, and a nutrient-rich site may tolerate a considerable nutrient removal. However, even on a nutrient-rich site, removal of nutrients without making sure they are replaced is inconsistent with the principles of sustainable forest management. Raulund-Rasmussen et al. (2008) suggested a nutrient balance approach for predicting sites at risk. This will require considerable knowledge of the various nutrient pools and fluxes (shown schematically in Fig. 1), which are sometimes difficult to obtain, leading to a large degree of uncertainty in nutrient balance calculations. A further approach suggested by Raulund-Rasmussen et al. (2008) was to classify forest soils into robust and sensitive types with respect to the risk for nutrient depletion (Table 1). Among relevant factors are temperature, soil depth, soil type (organic/mineral), soil texture, pH and mineralogy. In predicting site sensitivity, knowledge about similar sites is another useful tool. This knowledge can in many cases be obtained from literature studies, e.g. on harvesting experiments or fertilisation experiments.

To minimise the risk of nutrient depletion, it is important to develop methods for leaving the nutrient-rich foliage on site (Helmisaari et al., 2011). In forestry practice, piles of branches and tops are often left in the forest for periods of up to one year before removal, in order for as much as possible of the foliage to fall off (Fig. 2). This allows the return of the nutrients to the site.

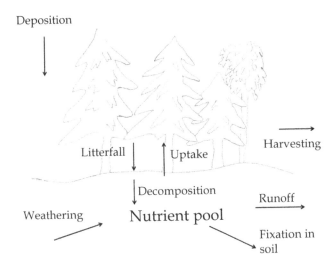

Fig. 1. Schematic overview of nutrient fluxes in the boreal forest ecosystem

Temperature	Depth	Type/texture		pH	Minerals	Sensitivity
<2°C	-	-	-	-	-	S
>2°C	<30 cm	-	-	-	-	S
>2°C	>30 cm	Organic	Fen	-	-	R/S
>2°C	>30 cm	Organic	Bog	-	-	S
>2°C	>30 cm	Mineral	Loamy	<4.8	-	S
>2°C	>30 cm	Mineral	Loamy	4.8-6	-	R
>2°C	>30 cm	Mineral	Loamy	>6	-	R
>2°C	>30 cm	Mineral	Sandy	<4.8	Quartz	S
>2°C	>30 cm	Mineral	Sandy	<4.8	Dark minerals	S
>2°C	>30 cm	Mineral	Sandy	4.8-6	Quartz	S
>2°C	>30 cm	Mineral	Sandy	4.8-6	Dark minerals	R
>2°C	>30 cm	Mineral	Sandy	>6		R

Table 1. Classification of soils into robust (R) and sensitive (S) types, based on Raulund-Rasmussen et al. (2008)

Fig. 2. Removal of a pile of branches and tops six months after harvesting, Gaupen, Norway (photo: Kjersti Holt Hanssen, Norwegian Forest and Landscape Institute)

There is some concern that piling of branches and tops might increase the risk for pest outbreaks, in contrast to direct removal of these residues after harvesting or chipping on-site. However, compared to SOH, piling, if carried out before insect colonisation, might even reduce the risk for outbreaks because larger amounts of wood (on the insides of the piles) would become less accessible to the pests (Schroeder, 2008). Piles of forest fuel from final cuttings are in some cases located close to stand edges of living trees, while whole trees from thinnings may be piled and stored in rows inside the stands; in these cases, there is a clear risk that bark beetles might attack nearby standing living trees (Schroeder 2008). However, it is not certain that the risk is great in practice. Recommendations include avoiding summer storage of large amounts of spruce with a diameter exceeding 10 cm close to mature living spruces, avoiding storage of spruce in thinned stands after warm and dry summers, and avoiding storage of both pine and spruce in defoliated forests (Schroeder 2008). National legislation regarding the amounts of coniferous wood that may be left or stored in the forest exists in many countries.

Returning wood ash to the forest has been suggested as a measure against nutrient loss, as all the major plant nutrients except nitrogen are found in wood ash. Wood ash input increases concentrations of base cations and reduces soil acidity (Arvidsson & Lundkvist, 2003; Brunner et al., 2004). Concentrations of potassium and magnesium in tree fine roots increase (Brunner et al., 2004). However, experiments with ash input on mineral soils have shown no significant increase (or decrease) in growth, probably due to nitrogen limitation (Karltun et al., 2008; Ozolinčius et al., 2007a). On peat soils, the situation is different: Swedish and Finnish experiments have shown an increase in tree growth after ash input (Karltun et al., 2008). There are concerns about increased concentrations of heavy metals after ash input (e.g. Reimann et al., 2008), especially in fungi and berries (Karltun et al., 2008). This will depend on the dose of ash added, and it is recommended to add a dose giving an amount of heavy metals no higher than the amount removed (Swedish Forest Agency, 2001). Effects on ground vegetation were limited when crushed hardened wood ash was used (Arvidsson et al., 2002). There may be a risk for damage to mosses and lichens (Ozolinčius et al., 2007b). Changes in mycorrhizal species composition have been observed (Karltun et al., 2008). The risk for negative effects appears to be low if the ash is treated before use, e.g. by hardening (Arvidsson et al., 2002) or added as granules (Callesen et al., 2007) or pellets (Rothpfeffer, 2007).

Soil organic matter (SOM) is an important reservoir for nutrients, especially nitrogen; its decomposition and mineralisation are important in nutrient cycling. In northern boreal forests, soil temperature and moisture are below optimal for decomposition, and changes in these after harvesting might be expected to increase decomposition and nutrient availability and leaching at least in the short term (Yin et al., 1989). However, increased soil moisture as a result of decreased evapotranspiration might lead to waterlogging and anaerobic conditions in the rooting zone that might inhibit decomposition (Prescott et al., 2000). In fact, the effect of harvesting on soil organic matter is variable. Decomposition rates of surface litter have been found to decrease after clear-cutting (Yanai et al., 2003), while accelerated mineralisation as a result of clear-cutting has been observed in Finland (Palviainen et al., 2004).

The effect of harvesting intensity on soil C and N has been found to vary greatly (Johnson and Curtis, 2001; Olsson et al., 1996a; Vesterdal et al., 2002). In their meta-analysis, Johnson and Curtis (2001) found different effects from different harvesting methods and tree species: SOH of coniferous species appeared to cause an increase in soil C while WTH caused a decrease. SOH of hardwoods, on the other hand, also appeared in most cases to lead to a decrease in soil C. A decrease in soil C was observed independent of harvest intensity for Norway spruce and Scots pine in Sweden (Olsson et al., 1996a). Harvest intensity may affect the decomposition of existing SOM as well as the build-up of new SOM from litter and forest residues.

At present, management recommendations for harvesting do not deal with optimisation of the carbon content of forest soils, although recommendations regarding erosion, soil compaction, drainage, and site preparation will clearly influence the carbon content. One reason for this is our incomplete understanding of the processes involved in carbon cycling in boreal forest ecosystems, and of which factors are most crucial for maximising carbon sequestration in these ecosystems.

Harvesting decreases evapotranspiration and thus increases runoff quantity. Haveraaen (1981) observed that clear-cutting might increase runoff by up to 40% in an area of eastern Norway with shallow soil. Harvesting also influenced water quality: nitrogen loss increased by up to six times (from 1.5 to 7-9 kg/ha), mostly (about 6 kg/ha) as nitrate. Corresponding increases were from 2 to 12-13 kg/ha for potassium, 18 to 24 kg/ha for sulphur in the form of sulphate, and from 16 to 35 kg/ha for chloride (Haveraaen, 1981). Removal of harvesting residues might possibly reduce runoff of these and other elements. Runoff water can become more acid after harvesting (Stupak et al., 2007).

Clear-cutting on Norway spruce-dominated drained peatlands has been shown to cause increased export of dissolved organic carbon (DOC) (Nieminen, 2004). Mineralisation of organic nitrogen followed by nitrification will increase nitrate concentrations. Because uptake is low, this nitrate will be largely leached from the system, together with base cations (Raulund-Rasmussen & Larsen, 1990). Nitrate in deposition will not be taken up to such a large extent as before harvesting, but will also be leached together with base cations. In Sweden, a clear nitrogen leaching gradient has been found on clear-cuts from the west to the east, following the deposition gradient but also influenced by higher runoff in the west (Akselsson et al., 2004). Increased export of all main forms of dissolved nitrogen (nitrate, ammonium and organic nitrogen) has been observed after harvesting (Haveraaen, 1981; Nieminen, 2004). However, small clear-cuts on a nitrogen-saturated site in Germany did not appear to increase the risk for nitrate contamination (Huber et al., 2004). P concentrations did not significantly increase, while P export increased only slightly after harvesting (Haveraaen, 1981; Nieminen, 2004). Base cation fluxes in runoff may increase after harvesting (Haveraaen, 1981; Hu, 2000), as increased decomposition of organic matter may lead to increased concentrations of base cations in runoff water. Piirainen et al. (2004) observed only slightly increased fluxes of P and base cations from below the B horizon after clear-cutting, despite increased fluxes from the O horizon.

Soil water chemistry in forest soils is affected by harvesting, with increased leaching of nutrients such as nitrogen and base cations after harvesting. For example, Hu (2000) found higher nitrate and potassium concentrations in soil water from mineral soils 2-3 years after harvesting and Piirainen et al. (2004) observed that the phosphorus flux under the organic layer increased three times and the base cation flux increased two times after SOH. This leaching is counteracted by growth, partly of ground vegetation (Fahey et al., 1991; Palviainen et al., 2005) and partly of new trees. Removal of forest residues influences soil water, as reduced concentrations of nitrate, ammonium and potassium have been observed (Staaf & Olsson, 1994). Where stumps had been removed, Staaf and Olsson found increased ammonium concentrations for two years, followed by two years of increased nitrate concentrations and acidity. These effects were only temporary: after four years there was no great difference between plots with stem-only harvesting, whole-tree harvesting, and whole-tree harvesting together with stump removal. As effects of harvesting on soil water chemistry change with time, it is important to have long-term experiments.

Significant forest resources are often located in more difficult situations, especially in mountain areas. Due to difficult access and the high cost of traditional (motor-manual) harvesting systems, these areas are currently underused. Today improved technical equipment as well as higher market prices make it possible to harvest also steeper slopes with partly- or fully-mechanized harvesting systems. Depending on the type of the technical system (wheeled or tracked harvesters, skidding/forwarding, cable systems etc.) and the design of the harvesting operation (distance and slope of the skid trails/roads) but also on soil quality and slope, various degrees of erosion and other physical damage to the soil can be observed after mechanized harvesting operations (Worrell & Hampson, 1997). Heavy erosion creates problems for soil, water and technical accessibility in the future. There is concern about the effect of increased sediment loads on water quality downstream of the harvested site: this might affect rural water treatment plants and fish reproduction (Nisbet, 2001). In addition, erosion causes loss of nutrients and organic matter from the forest ecosystem. Methods for reducing erosion risk are well-known, including for example building of culverts, bridges, and silt traps, and these methods have been incorporated in management guidelines in some countries (e.g. Forest Service, 2000; Forestry Commission, 2003).

3. Effects of harvesting intensity on biological diversity

Many organisms are dependent on logging residues as habitats or shelter, so removing this material for fuel will clearly affect these organisms' ability to survive. Species that depend on wood for their survival are termed saproxylic, and in northern Europe there exist several thousand such species, mainly fungi and insects. A further risk is that insects colonise wood bound for heating plants, and are thus trapped in wood that is burned. It is possible to make qualitative recommendations about which types of habitats or wood types that have the most threatened fauna and flora, based on information about landscape history and microhabitat associations   (Jonsell, 2007). For example, in Sweden, based on studies of saproxylic beetles, it appears that coniferous wood can be harvested as forest fuel to a rather large extent, whereas deciduous tree species, and especially southern deciduous species and aspen, should be retained to a larger degree (Jonsell, 2007). In addition to saproxylic species, other organisms which feed on them, such as woodpeckers, are also likely to be affected.

Some studies have dealt with effects on ground vegetation. Vegetation retains nutrients in the ecosystem and can decrease nutrient leaching prior to stand re-establishment after clear-cutting (Palviainen et al., 2005). WTH and removal of logging residues leads to reduced amounts of woody debris in clear-cuts and changes in physical and other environmental conditions (Åström et al., 2005), including soil nutrient contents (Staaf & Olsson, 1994), microclimate (Åström et al., 2005), increased light supply and mechanical disturbance. These changes could lead to changed species composition, reduced biodiversity and reduced nutrient content in the humus layer (Olsson et al., 1996a, 1996b). Increased biomass removal may change the abundance of plant species with a key ecosystem role (Bergstedt & Milberg, 2001). Differences in ground vegetation related to felling (selective vs. clear-cutting) have been found as long as 60 -70 years after harvesting (Økland et al., 2003).

Reported effects of increased biomass removal on boreal forest vegetation differ (e.g. Åström et al., 2005; Olsson & Staaf, 1995). Fahey et al. (1991) found that grass biomass increased more rapidly after WTH compared with SOH and continued to make up a higher proportion of the biomass during the first four years after harvesting, while

Bergquist et al. (1999) found no effects of WTH on grasses. Åström et al. (2005) found that WTH reduced bryophyte cover by half (hepatics in particular were affected) and increased graminoid cover with 10% but found no significant effects on other vascular plants, whereas Olsson and Staaf (1995) reported lower cover of most vascular plants after WTH, while bryophytes were unaffected by the logging method. These contrasting results may be due to several factors, e.g. differences in environmental and climatic conditions at the study sites, sampling methods and statistical treatment (T. Økland, personal communication).

4. Effects of harvesting intensity on forest regeneration and productivity

The major concern about WTH from the point of view of the forestry industry has been that the removal of branches and foliage will lead to reduced productivity in the next rotations, as these parts contain a large share of the nutrients in the tree. This has generally (although not always) been found to be the case. In Fennoscandia, Jacobson et al. (2000) demonstrated growth decreases in the first 10 years after WTH in thinnings of Norway spruce and Scots pine stands when compared with conventional thinnings. The growth reduction could be counteracted by nitrogen fertilisation and they concluded that the reduction was due to reduced nitrogen supply. The growth reduction continued in a second ten-year period, but could also then be compensated by fertilisation (Helmisaari et al., 2011). Results from one of the sites included by Jacobson et al. (2000) and Helmisaari et al. (2011), Bergermoen in Norway, are given in Table 2. In the Table, it can be clearly seen that plots where whole-tree thinning had been carried out (Treatment 2) had lower production than plots where stem-only thinning had been carried out (Treatment 1), and that this reduced production could be counteracted using fertilisation (Treatments 3 and 5).

Revision year	1	2	3	4	5
1989	168	163	177	180	182
1994	209	196	216	221	225
1999	255	239	259	265	261
2005	312	293	308	320	313

Table 2. Total production (m^3/ha) by treatment and revision year in an experiment with stem-only vs. whole-tree thinning with and without fertilisation in a Norway spruce forest at Bergermoen, Norway. Thinning took place in 1984. The treatments are: 1) stem-only thinning (SOT), 2) whole-tree thinning (WTT), 3) WTT + NPK compensation fertilisation, 4) SOT + 150 kg N + 30 kg P/ha, and 5) WTT + 150 kg N + 30 kg P/ha. The data are available at http://www.skogforsk.no/feltforsok/Langfig.cfm?Fnr=1057 (in Norwegian)

Egnell and Valinger (2003) also found reduced growth in a Scots pine stand 24 years after WTH as well as branch and stem harvest (BSH). Comparable results have been found in the UK by Proe and Dutch (1994) in second generation Sitka spruce after clear-cutting including removal of residues. However, the effect of WTH seems to be site-dependent as well as species-dependent, as not all studies have shown an unambiguous nutrient decrease with subsequent growth reduction after whole-tree harvesting (Egnell & Leijon, 1999; Olsson et al., 1996a). Results from the North American Long-Term Soil Productivity study showed only a limited effect of WTH compared to SOH: although growth decreased, seedling survival was in fact improved five years after WTH (Fleming et al., 2006).

5. Legislation, certification and management recommendations

As mentioned above, the increased use of renewable energy sources, including forest biomass, is a marked characteristic in current energy policy. In forest policy, the use of forest biomass for energy is usually supported as a sustainable form of energy that contributes to social welfare, rural development and the forest economy. Energy legislation is used directly as a tool to promote renewable energy including forest and other biomass, whereas forest legislation rather works to ensure sustainably produced forest biomass for all uses (Stupak et al., 2007). However, increased use of forest biomass for energy might lead to conflict between different interests, all of which are politically important: on the one side, the need for a secure and renewable source of energy as well as rural employment, and on the other ecologically sound long-term timber production, biological diversity and other uses of the forest such as recreation. Trade-offs between these various interests will then be necessary, and increased knowledge is essential in order to optimise these trade-offs. Sustainability principles and criteria have therefore to be incorporated into policy frameworks and support schemes, as well as management guidelines. Many countries have produced national recommendations and guidelines for forest fuel extraction and/or wood ash recycling to encourage the extraction of forest fuels taking place in agreement with the principles of sustainable forest management. Certification is another approach: the main forest certification schemes are the Programme for the Endorsement of Forest Certification schemes (PEFC) and Forest Stewardship Council (FSC). In national PEFC and FSC standards, issues related to wood for energy are included under several criteria. Recommendations elaborated by governments or other groups of stakeholders could be used for further development of legislation, certification standards, and guidelines in relation to the sustainable use of forest biomass for energy. Recommendations vary according to subject, but on the whole, economic, ecological and social questions are treated for the whole forest fuel chain, from removal of biomass from the forest to recycling of wood ash to the forest (Stupak et al., 2007). Scientific results must be interpreted and transferred to operational criteria, indicators, recommendations and guidelines, with the final thresholds being set by politicians, certification bodies or other stakeholders (Stupak et al., 2007). This interpretation will necessarily include a large degree of uncertainty, so that continuous further development will be necessary as new knowledge is obtained.

6. Conclusions

Removal of forest residues for bioenergy after harvesting might increase the risk for adverse effects on the environmental services provided by forests, such as water protection, carbon sequestration and biological diversity. Forest legislation, certification systems, and management guidelines have been developed in order to reduce the risk for non-sustainable use of forest resources. However, not enough is known at present about which factors determine the contrasting effects found in field experiments, and more research is therefore needed, and further development of legislation, certification standards and management guidelines is likely to take place as new knowledge is obtained.

7. Acknowledgement

This work was funded by the Research Council of Norway as part of Work Package 4.2 ("Ecosystem Management") of the Bioenergy Innovation Centre CenBio.

8. References

Akselsson, C.; Westling, O. & Örlander, G. (2004). Regional mapping of nitrogen leaching from clearcuts in southern Sweden. *Forest Ecology and Management*, Vol. 202, Nos. 1-3, (December 2004), pp. 235-243, ISSN 0378-1127

Arvidsson, H. & Lundkvist, H. (2003). Effects of crushed wood ash on soil chemistry in young Norway spruce stands. *Forest Ecology and Management*, Vol. 176, Nos. 1-3, (March 2003), pp. 121-132, ISSN 0378-1127

Arvidsson, H.; Vestin, T. & Lundkvist, H. (2002). Effects of crushed wood ash application on ground vegetation in young Norway spruce stands. *Forest Ecology and Management*, Vol. 161, Nos. 1-3, (May 2002), pp. 75-87, ISSN 0378-1127

Åström, M.; Dynesius., M.; Hylander, K. & Nilsson, C. (2005). Effects of slash harvest on bryophytes and vascular plants in southern boreal forest clear-cuts. *Journal of Applied Ecology*, Vol. 42, No. 6, (December 2005), pp. 1194-1202, ISSN 0021-8901

Bergquist, J.; Orlander, G. & Nilsson, U. (1999). Deer browsing and slash removal affect field vegetation on south Swedish clearcuts. *Forest Ecology and Management*, Vol. 115, Nos. 2-3, (March 1999), pp. 171-182, ISSN 0378-1127

Bergstedt, J. & Milberg, P. (2001). The impact of logging intensity on field-layer vegetation in Swedish boreal forests. *Forest Ecology and Management*, Vol. 154, No. 3, (December 2001), pp. 105-115, ISSN 0378-1127

Brunner, I.; Zimmermann, S.; Zingg, A. & Blaser, P. (2004). Wood-ash recycling affects forest soil and tree fine-root chemistry and reverses soil acidification. *Plant and Soil*, Vol. 267, Nos. 1-2, (December 2004), pp. 61-71, ISSN 0032-079X

Callesen, I.; Ingerslev, M. & Raulund-Rasmussen, K. (2007). Dissolution of granulated wood ash examined by in situ incubation: Effects of tree species and soil type. *Biomass and Bioenergy*, Vol. 31, No. 10 (October 2007), pp. 693-699, ISSN 0961-9534

Egnell, G. & Leijon, B. (1999). Survival and Growth of Planted Seedlings of *Pinus sylvestris* and *Picea abies* After Different Levels of Biomass Removal in Clear-felling. *Scandinavian Journal of Forest Research*, Vol. 14, No. 4, (July 1999), pp. 303-311, ISSN 0282-7581

Egnell, G., & Valinger, E. (2003). Survival, growth, and growth allocation of planted Scots pine trees after different levels of biomass removal in clear-felling. *Forest Ecology and Management*, Vol. 177, Nos. 1-3, (April 2003), pp. 65-74, ISSN 0378-1127.

Fahey, T.J.; Hill, M.O.; Stevens, P.A.; Hornung, M. & Rowland, P. (1991). Nutrient Accumulation in Vegetation Following Conventional and Whole-Tree Harvest of Sitka Spruce Plantations in North Wales. *Forestry* Vol. 64, No. 3, (July 1991), pp. 271-288, ISSN 0015-752X

Fleming, R.L.; Powers, R.F.; Foster, N.W.; Kranabetter, J.M.; Scott, D.A.; Ponder Jr., F.; Berch, S.; Chapman, W.K.; Kabzems, R.D.; Ludovici, K.H.; Morris, D.M.; Page-Dumroese, D.S.; Sanborn, P.T.; Sanchez, F. G..; Stone, D.M. & Tiarks, A.E. (2006) Effects of organic matter removal, soil compaction, and vegetation control on 5-year seedling performance: a regional comparison of Long-Term Soil Productivity sites. *Canadian Journal of Forest Research*, Vol. 36, No. 3, (March 2006), pp. 529-550, ISSN 0045-5067

Forest Service (2000). *Forest Harvesting and the Environment Guidelines*. Forest Service, Department of the Marine and Natural Resources, Dublin, Ireland, Available from http://www.agriculture.gov.ie/media/migration/forestry/publications/harvesting.pdf

Forestry Commission (2003). *Forests & Water Guidelines*. Forestry Commission, ISBN 0 85538 615 0, Edinburgh, U.K., Available from http://www.forestry.gov.uk/pdf/FCGL002.pdf/$FILE/FCGL002.pdf

Haveraaen, O. (1981). The effect of cutting on water quantity and water quality from an East-Norwegian coniferous forest. *Reports of the Norwegian Forest Research Institute*, Vol. 36, No. 7, pp. 1-27, ISSN 0332-5709 (in Norwegian with an English summary).

Helmisaari, H.-S.; Hanssen, K.H.; Jacobson, S.; Kukkola, M.; Luiro, M.; Saarsalmi, A.; Tamminen, P. & Tveite, B. (2011). Logging residue removal after thinning in Nordic boreal forests: Long-term impact on tree growth. *Forest Ecology and Management* Vol. 261, No. 11, (June 2011), pp. 1919-1927, ISSN 0378-1127

Hu, J. (2000). *Effects of harvesting coniferous stands on site nutrients, acidity and hydrology*. PhD Thesis, Department of Forest Sciences, Agricultural University of Norway, ISBN 82-575-0443-2, Ås, Norway

Huber, C.; Weis, W.; Baumgarten, M. & Göttlein, A. (2004). Spatial and temporal variation of seepage water chemistry after femel and small scale clear-cutting in a N-saturated Norway spruce stand. *Plant and Soil*, Vol. 267, Nos. 1-2, (December 2004), pp. 23-40, ISSN 0032-079X

Jacobson, S.; Kukkola, M.; Mälkönen, S. & Tveite, B. (2000). Impact of whole-tree harvesting and compensatory fertilization on growth of coniferous thinning stands. *Forest Ecology and Management*, Vol. 129, Nos. 1-3, (April 2000), pp. 41-51, ISSN 0378-1127

Johnson, D.W. & Curtis, P.S. (2001). Effects of forest management on soil C and N storage: meta analysis. *Forest Ecology and Management*, Vol. 140, Nos. 2-3, (January 2001), pp. 227-238, ISSN 0378-1127

Jonsell, M. (2007). Effects on biodiversity of forest fuel extraction, governed by processes working on a large scale. *Biomass and Bioenergy*, Vol. 31, No. 10, (October 2007), pp. 726-732, ISSN 0961-9534

Karltun, E.; Saarsalmi, A.; Ingerslev, M.; Mandre, M.; Gaitnieks, T.; Ozolinčius, R. & Varnagiryte, I. (2008). Wood ash recycling – possibilities and risks. In: *Sustainable use of forest biomass for energy – a synthesis with focus on the Nordic and Baltic countries*, D. Röser, A. Asikainen, K. Raulund-Rasmussen, I. Stupak I (Eds.), 79-108, Springer, ISBN 978-1-4020-5053-4, Heidelberg, Germany

Nieminen, M. (2004). Export of Dissolved Organic Carbon, Nitrogen and Phosphorus Following Clear-Cutting of Three Norway Spruce Forests Growing on Drained Peatlands in Southern Finland. *Silva Fennica* Vol. 38, No. 2, (April 2004), pp. 123-132, ISSN 0037-5330

Nisbet, T.R. (2001). The role of forest management in controlling diffuse pollution in UK forestry. *Forest Ecology and Management*, Vol. 143, Nos. 1-3, (April 2001), pp. 215-226, ISSN 0378-1127

Økland, T.; Rydgren, K.; Halvorsen Økland, R.; Storaunet, K.O. & Rolstad, J. (2003). Variation in environmental conditions, understorey species number, abundance and composition among natural and managed *Picea abies* forest stands. *Forest Ecology and Management* , Vol. 177, Nos. 1-3, (April 2003), pp. 17-37, ISSN 0378-1127

Olsson, B.A. & Staaf, H. (1995). Influence of Harvesting Intensity of Logging Residues on Ground Vegetation in Coniferous Forests. *Journal of Applied Ecology*, Vol. 32, No. 3, (August 1995), pp. 640-654, ISSN 0021-8901

Olsson, B.A.; Staaf, H.; Lundkvist, H.; Bengtson, J. & Rosén, K. (1996a). Carbon and nitrogen in coniferous forest soils after clear-felling and harvests of different intensity. *Forest Ecology and Management* , Vol. 82, Nos. 1-3, (April 1996), pp. 19-32, ISSN 0378-1127

Olsson, B.A.; Bengtsson, J. & Lundkvist, H. (1996b). Effects of different forest harvest intensities on the pools of exchangeable cations in coniferous forest soils. *Forest Ecology and Management* , Vol. 84, Nos. 1-3 (August 1996), pp. 135-147, ISSN 0378-1127

Ozolinčius, R.; Varnagirytė-Kabašinskienė, I.; Stakėnas, V. & Mikšys, V. (2007a). Effects of wood ash and nitrogen fertilization on Scots pine crown biomass. *Biomass and Bioenergy*, Vol. 31, No. 10, (October 2007), pp. 700-709, ISSN 0961-9534

Ozolinčius, R.; Buožytė, R. & Varnagirytė-Kabašinskienė, I. (2007b) Wood ash and nitrogen influence on ground vegetation cover and chemical composition. *Biomass and Bioenergy*, Vol. 31, No. 10, (October 2007), pp. 710-716, ISSN 0961-9534

Palviainen, M.; Finér, L.; Kurka, A.-M.; Mannerkoski, H.; Piirainen, S. & Starr, M. (2004). Release of potassium, calcium, iron and aluminium from Norway spruce, Scots pine and silver birch logging residues. *Plant and Soil*, Vol. 259, Nos. 1-2, (February 2004), pp. 123-136, ISSN 0032-079X

Palviainen, M.; Finér, L.; Mannerkoski, H.; Piirainen, S. & Starr, M. (2005). Changes in the above- and below-ground biomass and nutrient pools of ground vegetation after clear-cutting of a mixed boreal forest. *Plant and Soil*, Vol. 275, Nos. 1-2, (August 2005), pp. 157-167, ISSN 0032-079X

Piirainen, S.; Finér, L.; Mannerkoski, H. & Starr, M. (2004). Effects of forest clear-cutting on the sulphur, phosphorus and base cation fluxes through podzolic soil horizons. *Biogeochemistry*, Vol. 69, No. 3, (July 2004), pp. 405-424, ISSN 1573-515X

Prescott, C.E.; Blevins, L.L. & Staley, C.L. (2000). Effects of clear-cutting on decomposition rates of litter and forest floor on forests of British Columbia. *Canadian Journal of Forest Research*, Vol. 30, No. 11, (November 2000), pp. 1751-1757, ISSN 0045-5067

Proe, M.F. & Dutch, J. (1994). Impact of whole-tree harvesting on second rotation growth of Sitka spruce: the first 10 years. *Forest Ecology and Management*, Vol. 66, Nos. 1-3, (July 1994), pp. 39-54, ISSN 0378-1127

Raulund-Rasmussen, K. & Larsen, J.B. (1990). Cause and effects of soil acidification in forests – with special emphasis on the effect of air pollution and forest management. *DST - Dansk Skovbrugs Tidsskrift*, Vol. 75, pp. 1-41, (in Danish with an English summary), ISSN 0905-295X

Raulund-Rasmussen, K.; Stupak, I.; Clarke, N.; Callesen, I.; Helmisaari, H.-S.; Karltun, E. & Varnagiryte-Kabasinskiene, I. (2008). Effects of very intensive biomass harvesting on short and long term site productivity. In: *Sustainable use of forest biomass for energy – a synthesis with focus on the Nordic and Baltic countries*, D. Röser, A. Asikainen, K. Raulund-Rasmussen, I. Stupak (Eds.), 29-78, Springer, ISBN 978-1-4020-5053-4, Heidelberg, Germany

Reimann, C.; Ottesen, R.T., Andersson, M.; Arnoldussen, A., Koller, F. & Englmaier, P. (2008). Element levels in birch and spruce wood ashes-green energy? *Science of the Total Environment*, Vol. 393, Nos. 2-3, (April 2008), pp. 191-197, ISSN 0048-9697

Rothpfeffer, C. (2007). *From wood to waste and waste to wood – aspects on recycling waste products from the paper-pulp mill to the forest soil*. PhD Thesis, Department of Forest

Soils, Swedish University of Agricultural Sciences, ISBN 978-91-576-7382-4, Uppsala, Sweden

Schroeder, L.M. (2008). Insect pests and forest biomass for energy. In: *Sustainable use of forest biomass for energy – a synthesis with focus on the Nordic and Baltic countries*, D. Röser, A. Asikainen, K. Raulund-Rasmussen, I. Stupak (Eds.), 109-128, Springer, ISBN 978-1-4020-5053-4, Heidelberg, Germany

Staaf, H. & Olsson, B.A. (1994). Effects of slash removal and stump harvesting on soil water chemistry in a clearcutting in SW Sweden. *Scandinavian Journal of Forest Research*, Vol. 9, No. 4, (October 1994), pp. 305-310, ISSN 0282-7581

Stupak, I.; Asikainen, A.; Jonsell, M.; Karltun, E.; Lunnan, A.; Mizaraite, D.; Pasanen, K.; Pärn, H.; Raulund-Rasmussen, K.; Röser, D.; Schröder, M.; Varnagiryte, I.; Vilkriste, L.; Callesen, I.; Clarke, N.; Gaitnieks, T.; Ingerslev, M.; Mandre, M.; Ozolinčius, R.; Saarsalmi, A.; Armolaitis, K.; Helmisaari, H.-S.; Indriksons, A.; Kairiukstis, L.; Katzensteiner, K.; Kukkola, M.; Ots, K.; Ravn, H.P. & Tamminen, P. (2007). Sustainable utilisation of forest biomass for energy – possibilities and problems, policy, legislation, certification and recommendations. *Biomass and Bioenergy*, Vol. 31, No. 10, (October 2007), pp. 666-684, ISSN 0961-9534

Swedish Forest Agency (2001). Rekommendationer vid uttag av skogsbränsle och kompensationsgödsling. Meddelande 2/2001, Swedish Forest Agency, Jönköping, Sweden, ISSN 1100-0295 (in Swedish), Available at http://www.energiaskor.se/pdf-dokument/Kriterier/1518.pdf

Vesterdal L, Jørgensen FV, Callesen I, Raulund-Rasmussen K. 2002. Skovjordes kulstoflager – sammenligning med agerjorde og indflydelse af intensiveret biomasseudnyttelse. In: Christensen BT (Ed.), Biomasseudtag til energiformål – konsekvenser for jordens kulstofbalance i land- og skovbrug. DJF rapport Markbrug Vol. 72, (May 2002), pp. 14-28 (in Danish), ISSN 1397-9884

Worrell, R. & Hampson, F. (1997). The influence of some forest operations on the sustainable management of forest soils--a review. *Forestry*, Vol. 70, No. 1, (January 1997), pp. 61-85, ISSN 0015-752X

Yanai, R.D.; Currie, W.S. & Goodale, C.L. (2003). Soil Carbon Dynamics after Forest Harvest: An Ecosystem Paradigm Reconsidered. *Ecosystems*, Vol. 6, No. 3, (June 2003), pp. 197-212, ISSN 1432-9840

Yin, X.; Perry, J.A. & Dixon, R.K. (1989). Influence of canopy removal on oak forest floor decomposition. *Canadian Journal of Forest Research*, Vol. 19, No. 2, (February 1989), pp. 204-214, ISSN 0045-5067

Section 5

Biological Diversity

11

Ecological and Environmental Role of Deadwood in Managed and Unmanaged Forests

Alessandro Paletto[1], Fabrizio Ferretti[2], Isabella De Meo[1],
Paolo Cantiani[3] and Marco Focacci[4]

*[1]Agricultural Research Council – Forest Monitoring and
Planning Research Unit (CRA-MPF), Villazzano di Trento
[2]Agricultural Research Council – Apennine Forestry Research Unit (CRA-SFA), Isernia
[3]Agricultural Research Council – Research Centre for
Forest Ecology and Silviculture (CRA-SEL), Arezzo
[4]Land / Forestry Resources Consultant, Sesto Fiorentino
Italy*

1. Introduction

According to the Global Forest Resources Assessment 2005, forest deadwood encompasses all non-living woody biomass not contained in litter, either standing, lying on the ground, or in the soil (FAO, 2004). This definition considers the non-living biomass which remains in the forest regardless of the portion removed for production purposes (i.e. biomass-energy production). All different components of deadwood such as snags, standing dead trees (including high stumps), logs, lying dead trunks, fallen branches, fallen twigs and stumps are comprised in this account (Hagemann et al., 2009).

The role and importance of deadwood in forest ecosystem has been recognized by the international scientific community since the early 80's. The first scientific studies focused on the role of deadwood as a key factor for biodiversity conservation thanks to its ability in providing microhabitats for many species (Hunter, 1990; Raphael & White, 1984). Other issues were then discussed and analysed such as the protective role of coarse woody debris in stabilizing steep slopes and stream channels (Densmore et al., 2004), the contribution of deadwood to carbon, nitrogen and phosphorus cycles (Laiho & Prescott, 1999) and the influence on stand dynamics and regeneration of natural and semi-natural forests (Duvall & Grigal, 1999).

At the political level the recognition of its role was more recent, and raised in importance a decade after the scientific recognition. In particular, deadwood was included within the five carbon pool list (above-ground and below-ground biomass, litter, deadwood and soil) provided by Intergovernmental Panel on Climate Change (IPCC)-Good Practice Guidance for Land Use, Land Use Change and Forestry (IPCC-GPG) (2003). The change in C-stock in deadwood is required for reporting to the Kyoto Protocol (1997), Marrakesh Accords (7th Conference of the Parties, 2001) as well as to the United Nations Framework Convention on Climate Change (1992) (Tobin et al., 2007).

In Europe the importance of forest deadwood has been identified for the first time by the Ministerial Conference on the Protection of Forests in Europe (MCPFE) (2002) during the definitions of a set of Pan-European indicators for sustainable forest management. Deadwood is one of the indicators under the criterion "Maintenance, conservation and appropriate enhancement of biological diversity in forest ecosystems" and can be usefully considered in order to measure the level of biodiversity (Indicator 4.5: volume of standing deadwood and of lying dead-wood on forest and other wooded land classified by forest type).

The amount of deadwood in forest depends on a set of variables (Lombardi et al., 2008): forest type, stage of development, kind and frequency of natural or anthropogenic disturbances, local soil, local climatic characteristics and type of management. Regarding the latter, the qualitative and quantitative presence of deadwood in forest ecosystems is influenced by both forest system (either coppice or high forest) and the intensity of management (Green & Peterken, 1997; Fridman & Walheim, 2000). In managed forests, potential deadwood volumes are reduced by the extraction of timber and biomass. (Andersson & Hytteborn, 1991; Christensen et al., 2005; Green & Peterken 1997; Kirby et al., 1998; Verkerk et al., 2011)

Similarly, also the qualitative features are altered in comparison with those of natural dynamics. Furthermore, in traditional management practices, the accumulation of deadwood may not be desirable because of the increasing risk of either insect pests (such as bark beetles) or forest fires. In biodiversity oriented forest management one of the main purposes is to reduce the difference in deadwood volume between managed and unmanaged forests, while the close-to-nature forest management aims at maintaining a certain level of deadwood (Müller-Using & Bartsch, 2009).

For these reasons, in order to support technicians in developing suitable forest management plans and selecting the best silvicultural option, qualitative and quantitative data on deadwood should be collected. The silvicultural treatment can play a fairly significant role in balancing necromass volumes. Ad hoc solutions and well-designed planning of different silvicultural actions may increase, wherever is thought to be important, the presence of deadwood in the system. However, this achievement should be obtained without affecting costs and related management components (cost of harvesting, pest and fire hazard). With these premises, the authors provide a method to quantify stumps, standing and lying deadwood in forest with the aim at supporting multifunctional planning and management practices.

1.1 Functional and structural role of forest deadwood

Deadwood is an essential multifunctional and structural component of forest ecosystems (Harmon et al., 1986), being a key factor in the nutrient cycle (N, P, Ca and Mg) (Holub et al., 2001), a fundamental element in the ecological, geomorphological and soil hydrological processes (Bragg & Kershner, 1999), a relevant forest carbon pool (Krankina & Harmon, 1995), a potential resource for biomass-energy production and an important habitat for many species (mammals, birds, amphibians, insects, fungi, moss and lichen communities) (Nordén et al., 2004).

From the ecological and environmental point of view, deadwood increases the structural and biological diversity of the ecosystem since many organisms are adapted to utilise this resource. In particular, two types of organisms, which depend on its presence in the forest ecosystem, can be distinguished (Wolynski, 2001):

- directly dependent organisms: use the deadwood as substrate for germination, power source and nesting site;
- indirectly dependent organisms: find occasionally shelter in the coarse woody debris or in standing dead trees at either first or advanced level of decomposition.

Among the organisms of the first group, saproxylic insects occupy the most important position, being a major part of biodiversity in forest ecosystem (Schlaghamersky, 2003). The saproxylic organisms, either those classified as obligatory or facultative, depend, at some stage of its life cycle, on deadwood of senescent trees or fallen timber.

Considering the relationship between deadwood and bird species, a particularly important role is played by the "habitat trees". Normally, "habitat trees" are large size individuals, with a diameter greater than 30 cm, which contain hollows used by forest fauna (Humphrey et al., 2004). The bird species that are hosted by dead trees can be primary excavators of cavities (i.e. wood peckers) or secondary cavity nesters (Hagan & Grove, 1999). The importance of deadwood as an indicator of biodiversity is provided by the diameter of the tree which is closely related, in turn, to the size of the nest holes. Thus, some bird species, such as *Parus palustris, Parus caeruleus, Passer montanus,* and *Sitta europaea* require small cavities (hole diameter less than 5 cm), whereas some other species such as *Strix aluco, Upupa epops* (Longo, 2003), *Dryocopus martius, Picoides leucotos* and *Picoides major* need larger cavities (hole diameter more than 5 cm).

Moreover, several mammal species use hollows, cavities, roots, fallen branches and deadwood such as bear, lynx, fox, martens, squirrels, bats and many small rodents (Radu, 2006). In particular, many Mustelids use the deadwood as a shelter: the stone marten (*Martes foina*), marten (*Martes martes*) and wolverine (*Gulo gulo*). The tree holes are also used by common dormouse (*Muscardinus avellanarius*) and fat dormouse (*Myoxus glis*) as nesting site (Paolucci, 2003).

Deadwood plays also a role in soil stabilization, since the lying logs on the soil surface control the movements of soil and litter across the ground surface (Harmon et al., 1986). With special regard to the protective function of forests (indirect protection), the fallen logs may retain soil and water movement either on slopes or through the ground (Kraigher et al., 2002). Similarly, the fallen tree trunks provide good protection against avalanches and rockfalls (direct protection - Berretti et al., 2007). Considering the latter, lying deadwood has a positive effect in the short-medium term, whereas the decomposition of the wood can bring back the movement of stones accumulated over time.

Moreover, deadwood can act as a temporary storage site for carbon (C), because of its slow carbon dioxide release ability, thereby showing a potential in moderating global warming (Keller et al., 2004). Deadwood, as a carbon pool, can account for a substantial fraction of stored carbon, but only few studies have provided quantitative features and the length of the turnover in comparison with other C storing components of the forest ecosystem (above-ground biomass, below-ground biomass, litter and soil organic C) (Kueppers et al., 2004). The few efforts on this topic show that standing and lying deadwood accounts for about 6% of total carbon stock in forest (Ravindranath & Ostwald, 2008), but, according to a set of qualitative and quantitative features, it does so with a certain variability.

1.2 Quantitative and qualitative features of forest deadwood

The importance of deadwood in forest ecosystems can be analysed by considering some qualitative and quantitative features such as: volume and its distribution by component and size, origin (species or botanical group), decay class and spatial distribution.

The main variable to be considered, in order to evaluate and analyze the ecological importance in forests, and its influence on other ecosystem components, is volume. Volume was measured during the forest inventory by applying two main procedures (Morelli et al., 2006): the line transect method (Line Intersect Sampling - LIS), which was applied in order to quantify directly the amount of deadwood on the ground (Van Wagner, 1968) and the measurement of the metric attributes (length and diameter) in ordinary sample plots, in order to calculate both standing and lying deadwood volumes (Harmon & Sexton, 1996).

As indicated in literature, deadwood volumes vary greatly in forest: unmanaged natural or semi-natural forests show the highest values with more than 100 m^3 ha^{-1} (Green & Peterken, 1997), while intensively managed forests manifested the lowest outputs with 5-30 m^3 ha^{-1} (Kirby et al., 1998). This variability is influenced by the classification of components sizes as well as the diametric threshold used in the inventory and the different measurement methods. These data are confirmed by the results of the National Forest Inventories (NFIs). As explained in Table 1, the different management traditions existing in Europe have a direct consequence in the accumulation of deadwood. The highest values are recorded in central European forests (Austria, Germany and Switzerland), whereas the lowest values are found in France and Finland. In Italy the volume of all deadwood components (stump, standing and lying deadwood) amounts to 8.8 m^3 ha^{-1} (INFC, 2009).

At local level, different situations can be found since potential deadwood volumes are reduced in managed forests either by the extraction of timber and biomass (Verkerk et al., 2011) or by sanitation cuttings. Also the qualitative features are altered in comparison with those of natural dynamics (Hodge & Peterken, 1998). In the traditional forest management the presence of deadwood is considered negatively for several reasons. Historically, deadwood has been removed in order to decrease firewood risk as well as to protect timber from insect and fungal attacks (Radu, 2006). Nowadays, the newest paradigms in forest management have recognized the ecological role of deadwood and developed strategies to both maintain or increase the amount of deadwood in forest ecosystem (i.e. by increasing volume of lying deadwood in order to favour population of invertebrates).

Country	Volume (m^3 ha^{-1})	Note
Austria	13.9	NFI 2000-2002 Threshold considered 20 cm
Belgium	9.1	Standing and lying deadwood of Wallonia region
Finland	5.6	NFI 1996-2003. Threshold considered 10 cm
France	2.2	NFI 2002. Threshold considered 7.0 cm
Germany	11.5	NFI 2001-2002. Threshold considered 20 cm
Italy	8.8	NFI 2005. Threshold considered 10 cm
Norway	6.8	Threshold 10 cm
Sweden	6.1	NFI 1993-2002. Threshold considered 10 cm
Switzerland	11.9	NFI 1993-199. Threshold considered 7.0 cm

Source: Brassel & Brändli, 1999; Fridman & Walheim, 2000; INFC, 2009; Mehrani-Mylany & Hauk, 2004; Pignatti et al., 2009; Vallauri et al., 2003.

Table 1. Volume of deadwood in the main European Forest Inventories

Deadwood can be subdivided into three main components (Næsset, 1999): (1) snags or standing dead trees, (2) logs or lying deadwood and (3) stumps. All of them occupy a different ecological role in the forest ecosystem. Standing dead trees play an important role in increasing natural diversity and, in general, in the functioning of forest ecosystems, since a wide number of plants and animals has been strongly associated with their presence (Marage & Lemperiere, 2005). Lying deadwood provides important habitats for numerous insect species including flies, beetles and millipedes, while nurse logs facilitate the germination of conifers in mountain forests (Vallauri et al., 2003). Referring to the origin of deadwood, each piece can be classified on the basis of the species or by simply distinguishing between coniferous or deciduous. The tree species can be easily identified if the plant has recently died by observing the bark and the wood structure. When these parameters are no longer recognizable because of the advanced state of decomposition, a simple distinction conifer/broadleaved should be applied (Stokland et al., 2004).

Decomposition is the process through which the complex organic structure of biological material is reduced to its mineral form, and it is the result of the interactions between biotic and abiotic factors such as non-enzymatic chemical reactions, leaching, volatilisation, comminution and catabolism. Decay processes depend on species (hardwood or softwood species), site conditions (microclimate) and exposure; these characteristics can be quantified visually by using a decay class scale. The decay rate influences the dynamics of carbon release and sequestration, and it is measured with a decay class scale that takes into account species, microclimate and exposure. Moreover, the stage of decay is a very important parameter in order to analyse ecological dynamics and quantify carbon pools (Zell et al., 2009). Several ways to classify the level of decomposition can be found in literature. Normally the most widely accepted classification considers different (three, four or five) decay classes determined on the basis of the following variables (Montes et al., 2004): structure of bark, presence of small branches, softness of wood and other visible characteristics. The most common classification system is a 5-class system (Hunter, 1990) used in the American Forest Inventory (Waddell, 2002) and in the main European forest inventories (Paletto & Tosi, 2010; Sandström et al., 2007). The five decay classes used in the international standard are reported in Table 2.

Decay class	Bark condition	Small branches	Woody consistency	Other visual characteristics
1- Recently dead	Entire and attached	Present	Intact	Little rotten area under bark
2- Weakly decayed	Entire but not-attached	Partly present	Intact	Rotten areas < 3 cm
3- Medium decayed	Fragments of bark	Absent	Partly broken	Rotten area > 3 cm
4- Very decayed	Absent	Absent	Broken	Large rotten area
5- Almost decomposed	Absent	Absent	Dust	Very large rotten area, musk and lichens

Table 2. Decay classes of deadwood (five-class system)

Considering the size, lying deadwood is normally divided into two categories: coarse woody debris (CWD) which includes the logs with minimum diameter of 10 cm and fine woody debris (FWD) which refers to logs smaller than this threshold (Densmore et al., 2004). The same values are used to classify standing dead tree and stumps. Woldendorp et al. (2002) suggested to consider litter those small woody debris which have diameters below

2.5 cm, whereas other authors classify them as very fine woody debris (VFWD – Hegetschweiler et al., 2009). The distinction between these two (or three) categories of size is important when the habitat requirements of animal, fungi and plant species must be identified. Normally, FWD is relevant for diversity of wood-inhabiting fungi, especially ascomycetes in boreal forest whereas CWD favours many species of basidiomycetes. VFWD, in particular, may be associated to wood-inhabiting basidiomycetes, especially in managed forests where there is little availability of other substrates (Küffer & Senn-Irlet, 2005).

The spatial density, as a parameter, (Comiti et al., 2006) indicates how CWD and FWD are distributed on the ground. The spatial distribution is the result of human activities (i.e. cutting) or natural events (i.e. landslides). This variable can be qualitatively divided into three classes: (1) homogeneously concentrated, (2) concentrated in small groups, (3) scattered.

2. Materials and methods

The quantitative and qualitative features of snags, stumps and logs were estimated and analysed according to the forest types and forest systems (coppice and high forest) in three case studies. Hence, the authors examined the relationship between the presence of deadwood - species, size and decay class distribution - and the forest management practices.

The analysis of the influence of forest management on deadwood were investigated in a four-phases research:

- Classification of land uses/cover;
- Qualitative and quantitative description of forest formations;
- Dendrometric measures including the qualitative and quantitative information on forest deadwood;
- Analysis of the relationship between carbon storage and intensity of management.

The CORINE land cover (EC, 1993) European classification - level III - was adopted as a reference classification system for basic cartography. A specific classification was assembled for the forests that was based on the use of a homogeneous cultivation subcategory. This feature was ranked as an intermediate between the forest category and the forest type, and took into account both the forest system and the possible treatments of the wood. This classification was obtained according to the existing regional forest types and was coherent with higher superior reference systems (Italian National Forest Inventory-INFC, European Nature Information System - EUNIS, CORINE).

Regarding the qualitative and quantitative description of forest formations (woodlands and shrub lands), stratified samplings were conducted on the basis of homogeneous cultivation subcategory. The information was then entered in a Geographical Information System (GIS) built on the regional forest map.

The main dendrometric and management parameters were calculated in a sub-sample of woodlands using a circular area with a radius of 13 m measured onto a topographic map for a total surface of 531 m². The parameters measured were: number of trees, diameter at breast height (dbh), tree height of some sample trees, regeneration, deadwood, and qualitative attributes linked to the forest management and harvesting operations.

This method was tested on three study areas (forest districts) located in three different administrative regions of Southern Italy (Figure 1): (1) Arci-Grighine district in Sardinia region, (2) Alto Agri district in Basilicata region and (3) Matese district in Molise region.

The Arci-Grighine district (39°42'7" N; 8°42'4" E) is located in the Centre-East area of the Sardinia island. The district has a total surface of 55,183 ha and a population of 26,207 inhabitants (density of about 0.47 persons ha^{-1}) subdivided in 21 municipalities. The forests cover a surface area of 26,541 ha, comprising 48.1% of the Arci-Grighine territory. The forest categories, in order of prevalence, include: Mediterranean maquis (57.0%), Mediterranean Evergreen Oak forests (*Quercus ilex* L.) (14.7%), Cork Oak forests (*Quercus suber* L.) (9.9%), Eucalyptus forests (*Eucalyptus spp.*) (5.3%), Mediterranean pine forests (*Pinus spp.*) (3.1%), Downy oak forests (*Quercus pubescens* Willd.) (2.3%) and Monterey Pine (*Pinus radiata* Don) (1.7%)

The Alto Agri district (40°20'25" N; 15°53'52" E) is located in the Province of Potenza and characterized by a population of 33,739 people and a surface of 72,469 ha (density of about 0.47 persons ha^{-1}) divided into 12 municipalities. The forest areas cover 42,367 ha, comprising 58.4% of the Alto Agri territory. The forest categories, in order of prevalence, include: Downy oak forests (*Quercus pubescens* Willd.) (28.4%), followed by Turkey oak (*Quercus cerris* L.) forests (17.8%), shrub lands such as broom thicket, mixed thorny thicket and thermophile thicket with *Phillyrea sp.* and *Pistacia lentiscus*, (12.7%) and Beech (*Fagus sylvatica* L.) forests (9.6%).

The Matese district (41°29'12" N; 14°28'26" E) is located in the Province of Campobasso and characterized by a population of 21,022 people and a surface area of 36,539 ha (density of about 0.58 persons ha^{-1}) divided into 11 municipalities. The forest areas cover 15,712 ha, comprising 43.0% of the Matese territory. The forest categories, in order of prevalence, include: Turkey oak (*Quercus cerris* L.) forests (42.3%), followed by Beech (*Fagus sylvatica* L.) forests (30.5%), Hop hornbeam (*Ostrya carpinifolia* Scop.) forests (10.9%).

Fig. 1. Location of the case studies in Italy

The number of sub-plots were proportionally chosen according to the different forest surfaces: 218 sub-plots in Arci-Grighine, 235 sub-plots in Alto Agri and 117 sub-plots in Matese.

The quantitative presence of forest deadwood (volume) was investigated in each sub-sample plot taking into account four main integrative features: components, origin, decay class and size.

The volume of each log or snag included in the sub-sample was measured by applying a geometric system and, only for the snags, the stereometric equation of Italian National Forest Inventory 1985.

The forest operators registered lengths and diameters in three cross sections (minimum, maximum and medium) for lying dead wood while for the standing dead trees also the tree height and diameters at breast height (dbh) were considered.

Standing dead tree volume (V_s) was calculated from stand basal area (BA) whereas tree height was obtained from the hypsometric curve (h), by using the standard biometric equation (Cannell, 1984):

$$V_s = f \cdot BA \cdot h \tag{1}$$

which includes a standard stem form factor (f) of 0.5.

Lying deadwood volume (V_l) and stump volume (V_{st}) was calculated using the following formula:

$$V = \frac{\pi}{4} \cdot h \cdot \frac{D+d}{2} \tag{2}$$

Where:
h = height or length measured (m)
D = maximum diameter (m)
d = minimum diameter (m)
The total volume of deadwood in forest (V_d) was the sum of three components:

$$V_d = V_s + V_l + V_{st} \tag{3}$$

3. Results and discussion

3.1 Volume by components and decay class

The quantitative data on deadwood (snags, logs and stumps) in the three districts showed interesting differences linked to the different traditions of management (Table 3 & Figure 2). The maximum value of deadwood was found in the Matese district with 47.1 m³ ha⁻¹ being stumps (30.4 m³ ha⁻¹) and snags (9.9 m³ ha⁻¹) the major contributors. In the Arci-Grighine district the total volume was 21.2 m³ ha⁻¹, almost exclusively concentred in snags (19.2 m³ ha⁻¹). The Alto Agri district showed the lowest volumes of deadwood (8.8 m³ ha⁻¹), the majority (61%) comprised of snags (5.4 m³ ha⁻¹).

The variable number of deadwood pieces and volume provided an average of 0.54 m³ piece⁻¹, with a minimum value in the Alto Agri district (0.23 m³ piece⁻¹) and a maximum value in the Matese district (0.90 m³ piece⁻¹).

The results obtained were also compared with the Italian NFI. The volumes in the three districts were higher than those provided by NFI (INFC, 2009). In addition, the quantitative and qualitative differences were found to be comparable. A total of 1.7 m³ ha⁻¹ were recorded in Sardinia (0.8 m³ ha⁻¹ snag, 0.4 m³ ha⁻¹ stump and 0.5 m³ ha⁻¹ of log), 2.2 m³ ha⁻¹ in Basilicata (1.1 m³ ha⁻¹ snag, 0.5 m³ ha⁻¹ stump and 0.6 m³ ha⁻¹ log) and 4.3 m³ ha⁻¹ in Molise (2.7 m³ ha⁻¹ snag, 1.0 m³ ha⁻¹ stump and 0.6 m³ ha⁻¹ log). Other studies conducted in Italy show various values: 71.3 m³ ha⁻¹ were estimated in a site of Basilicata (Cozzo Ferriero) and in three sites of Molise of 95.6 m³ ha⁻¹ (Abeti Soprani), 17.4 m³ ha⁻¹ (Collemelluccio) and 26.5 m³ ha⁻¹ (Monte di Mezzo) (Lombardi et al., 2010). Moreover, in 21 study areas in North-West of Molise, Marchetti and Lombardi (2006) measured 15.1 m³ ha⁻¹. These data show how the great variability in volumes is associated with specific site conditions and management practices.

District	Snag		Log		Stump		Total
	N ha⁻¹	Volume (m³ ha⁻¹)	N ha⁻¹	Volume (m³ ha⁻¹)	N ha⁻¹	Volume (m³ ha⁻¹)	Volume (m³ ha⁻¹)
Alto Agri	11.3	5.4	16.3	3.3	11.0	0.1	8.8
Arci-Grighine	17.1	19.2	7.3	1.2	17.9	0.9	21.3
Matese	16.2	9.9	20.4	6.8	15.8	30.4	47.1
Mean	*14.9*	*11.5*	*14.7*	*3.8*	*14.9*	*10.5*	*25.7*
St.dev.	*3.1*	*7.0*	*6.7*	*2.8*	*3.5*	*17.3*	*19.5*

Table 3. Volume and number of pieces for the different components of deadwood by district

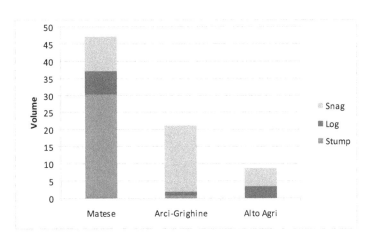

Fig. 2. Distribution of deadwood components volume (m³ ha⁻¹) by district

The variation in decay class distribution provides an indication of the temporal variation in both tree mortality and tree felling and this variable can be used as an indicator of the history of a forest (Rouvinen et al., 2005). Generally, when fallen dead trees show all decay classes, the death of the plants have probably occurred evenly over a long time. *Vice versa*, when decay stages are concentrated in one or few classes, external events (naturally or human-induced) have concentrated the mortality in specific moments. Two different situations were observed in the case studies (Figure 3). In the Arci-Grighine district the volume of deadwood was concentrated in the first decay class (around 80% of total deadwood) and was composed almost exclusively of standing deadwood. Probably, the dead material might have been deliberately left in the forest for ecological or economic reasons after recent cuttings. Conversely, deadwood was regularly distributed along the five decay classes in the other two districts. The Matese scored higher values in the strongly decayed classes (fourth and fifth class) with around 67% of total volume, while the Alto Agri showed an opposite trend with 76% of the volume concentrated in the first two classes.

In general, the relatively scarce presence of highly decayed material in the Alto Agri and Arci-Grighine districts may be related to the effect of repeated clearing of the undercover vegetation which was carried out in order to prevent forest fires. Furthermore, heavy forest grazing in the Alto Agri caused the removal of dead material in the past.

The difference distribution of deadwood volume by decay classes in the three case studies were compared using the Kruskal-Wallis non-parametric test (Table 4). The results showed statistically significant differences only for stumps. In particular, the differences were related to the different distribution of deadwood in the Matese district compared with the other two districts.

	Observed K	Critical value	Degrees of freedom	p-value	α
Snag	0.081	5.991	2	0.961	0.05
Log	1.940	5.991	2	0.379	0.05
Stump	12.020	5.991	2	0.002	0.05

Table 4. Kruskal-Wallis non-parametric test: difference among the three districts concerning the three deadwood components

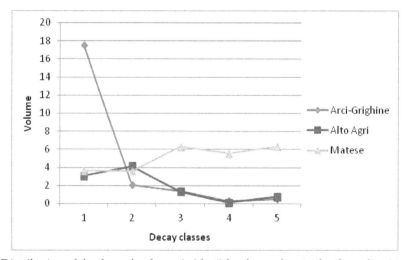

Fig. 3. Distribution of deadwood volume (m³ ha⁻¹) by decay class in the three districts

3.2 Diametric and species distribution

The diametric distribution of deadwood (Figures 4, 5 & 6) provides important information on the presence of habitat trees and the differences among the three components of deadwood itself.

Regarding the tree habitat, a minimum number of 5-10 trees ha^{-1} is required for biodiversity conservation, especially for saproxylic organisms (Mason et al., 2005). The situation found in the three districts varied greatly since in the Arci-Grighine 31 dead trees ha^{-1} and 44 logs ha^{-1} with a minimum diameter of 30 cm were recorded, whereas both in Alto Agri and Matese habitat trees were only 1.7 per hectares.

In addition, the results on diametric distribution showed two different situations, being the Alto Agri more represented in small diameter classes and the Arci-Grighine and Matese in high diameter classes. In particular, in the Alto Agri district all three components were concentred in the first diametric class (40.4% of snags and 58.6% of logs). Instead, in the Arci-Grighine around 36% of snags and 41% of logs and stumps fell above the 30 cm diameter class. Similarly, in the Matese district 51% of logs and 69% of stumps were distributed in the highest diametrical class. Probably, the Alto Agri differed so significantly from the other two districts because almost 30% of its forests is constituted of young evergreen oak coppices, with high densities and a continuous mortality of thin dominated individuals.

The difference between the diametric distribution of the deadwood components in the three case studies were compared in pairs through the use of Kolmogorov-Smirnov nonparametric test. This test is based on the difference in the cumulative distributions of the two datasets. The results showed in all cases no statistically significant differences.

In order to test these differences by forest district, the Chi-square test (χ^2) was applied to the three deadwood components. The results obtained (Table 5) showed a statistical difference in sampling distribution of stumps and logs, while for the snag distribution the difference among the three districts was not significant.

	Observed chi-square value	Calculated chi-square value	Degrees of freedom	p-value	α
Snag	11.738	15.507	8	=0.163	0.05
Log	94.668	15.507	8	< 0.0001	0.05
Stump	51.274	15.507	8	< 0.0001	0.05

Table 5. Chi-square test (χ^2): difference among the three districts concerning the three deadwood components

Regarding the deadwood distribution per species in the Arci-Grighine district, a total of 60% of non-living biomass was concentrated in a single species (Monterey pine). Similarly, in the Matese district 64% and 9.7% of deadwood belonged to European beech and Turkey oak respectively. These results may be explained by the active firewood collection in oak forests and the substantial abandonment of beech forests. The species in the Alto Agri district were more evenly distributed: 34.4% of deadwood belonged to European black pine, 15.1% to chestnut and 12.2% to European beech.

In the Matese district, the beech deadwood consisted mainly of stumps (45.9%) and logs (48.7%), probably originated by old cuttings. In the Arci-Grighine instead, the presence of Monterey pine deadwood was almost exclusively composed by standing trees coming from abandoned old plantations.

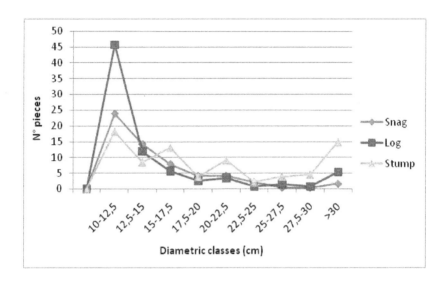

Fig. 4. Diametric distribution of deadwood in Alto Agri district

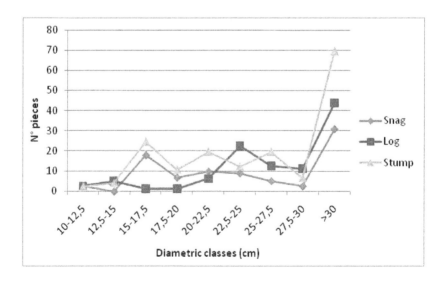

Fig. 5. Diametric distribution of deadwood in Arci-Grighine district

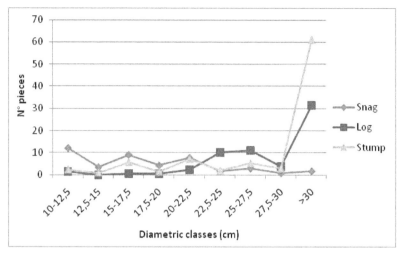

Fig. 6. Diametric distribution of deadwood in Matese district

3.3 Forest type and forest system

The type of forest system applied to the forest is a key factor to understand the impacts of forest management on deadwood and, consequently, on biodiversity conservation. The results showed that high forests had on the average higher volumes of deadwood for all three components in comparison with coppices (Table 6). In particular, the greatest differences were found for stumps in the Matese district (coppice: 7.4 m³ ha-1, high forest: 53.1 m³ ha-1) and for snags in the Arci-Grighine district (coppice: 3.8 m³ ha-1, high forest: 25.9 m³ ha-1). Only in the Matese district, snags scored higher values in coppices (11.1 m³ ha-1) rather than in high forests (8.7 m³ ha-1). This result was probably caused by a higher number of abandoned coppices in the area.

	Matese		Arci-Grighine		Alto Agri	
	Coppice	High forest	Coppice	High forest	Coppice	High forest
Stump	7.4	53.1	0.8	0.9	0.1	0.2
Log	4.4	9.1	0.2	1.6	1.6	5.9
Snag	11.1	8.7	3.8	25.9	3.9	7.6

Table 6. Volume (m³ ha-1) of deadwood components by forest system

Regarding the forest type, interesting differences were retrieved: in the Matese district *Fagus sylvatica* forests showed higher values than those of *Quercus cerris* forests, except for snags (Figure 7). In the Arci-Grighine district (Figure 8) very high volumes of snags were recorded in two forest type: *Pinus radiata* forests (93.5 m³ ha-1) and *Eucalyptus sp.* forests (54.1 m³ ha-1). In the Alto Agri district, instead, (Figure 9) Mediterranean pine forests (29.5 m³ ha-1) and, secondly, Mixed broadleaved forests (12.7 m³ ha-1) and *Castanea sativa* forests (12.8 m³ ha-1) showed the highest values of deadwood.

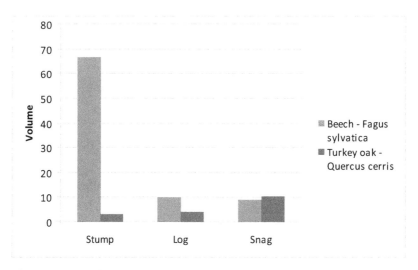

Fig. 7. Volume (m³ ha⁻¹) distribution per forest type in the Matese district

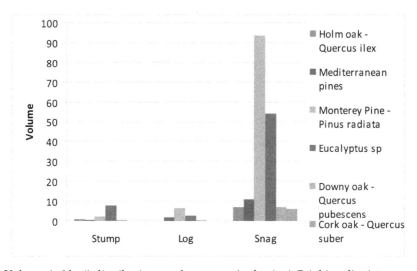

Fig. 8. Volume (m³ ha⁻¹) distribution per forest type in the Arci-Grighine district

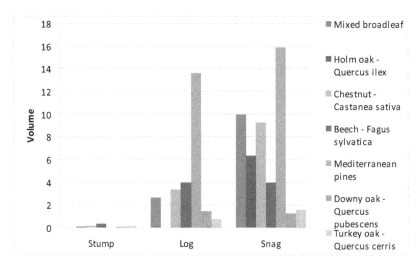

Fig. 9. Volume (m³ ha⁻¹) distribution per forest type in the Alto Agri district

The higher number of stumps in the beech forests in comparison with those of the Turkey oak forests in the Matese district was a direct consequence of the different silvicultural treatments. As a matter of fact, these residues in the *Fagus sylvatica* forests, which belonged mainly to old standard trees, were originated by the conversion to high forests of abandoned coppices.

In addition, these types of forest had traditionally undergone to a less active management because of a minor economic interest on its main product (firewood) and a generally difficult accessibility. Similar considerations explain the high number of snags in the Arci Grighine *Pinus radiata* and *Eucaliptus spp.* forests. The abandonment, in the last decades, of these plantations increased the competitions among the individuals, thereby promoting high mortality rates and big sized deadwood material.

4. Conclusions

A method to collect quantitative and qualitative features of deadwood was a useful tool in order to define management strategies and silvicultural treatments aimed at optimizing the presence of deadwood in forest. In addition, its importance is remarked by the relevance of deadwood in carbon sequestration.

The expeditious method for the quantification of deadwood has an effective management relevance in supporting the choice of silvicultural treatments for the different forest type. The planners, with the analysis of the deadwood stock distinguished by type, specie and modality of active management, may acquire fundamental elements in order to define the sustainability of their technical proposals. Hence, appropriate interventions, aimed at valorising the specific functions of deadwood, can be defined case by case.

The different techniques may prescribe, wherever necessary, either the release of standing dead trees or other particular actions to increase deadwood. In coppices, for instance, a few standards may be left to indefinite ageing as well as some declining or dying individuals may be chosen as standards. In high forests instead, snags may be artificially increased by girdling some plants. The number of stumps and logs may be improved by releasing dominated plants without economic value that will rapidly die and fall down. However, the increase of deadwood should be carefully planned along with all the remaining management considerations such as production, protection etc, giving particular attention on fire hazard and pest control.

5. Acknowledgments

Funding for this project was provided by INEA Basilicata, Regione Molise and Regione Sardegna. The authors contributed equally to this work.

6. References

Andersson, L.I. & Hytteborn H. (1991). Bryophites and decaying wood. – a comparison between managed and natural forest, *Ecography*, Vol. 14, pp. 121-130, ISSN 1600-0587

Berretti, R., Caffo, L., Camerino, P., De Ferrari, F., Domaine, A., Dotta, A., Gottero, F., Haudemand, J.C., Letey, C., Meloni, F., Motta, R. & Terzuolo, P.G. (2007). Selvicoltura nelle foreste di protezione. *Sherwood*, Vol.134, No.6, pp. 11-38, ISSN 1590-7805.

Bragg, D.C., Kershner, J.L. (1999). Coarse Woody Debris in Riparian Zones. *Journal of Forestry*, Vol.97, No.4, pp. 30-35, ISSN 0022-1201.

Brassel, P., Brändli, U.B. (1999). *Inventario forestale nazionale svizzero. Risultati del secondo inventario 1993-1995*. Birmensdorf, Istituto federale di ricerca per la foresta, la neve e il paesaggio. Haupt, Berna, Stoccarda and Vienna.

Cannell, M.G.R. (1984). Woody biomass of forest stands. *Forest Ecology and Management*, No.8, pp. 299-312, ISSN 0378-1127.

Christensen, M., Hahn, K., Mountford, E.P., Standovar, T., Rozembergar, D., Diaci, J., Wiildeven, S., Meyer, P., Winter, S. & Vrska, T. (2005). Dead wood in European beech (*Fagus sylvatica*) forest reserves. *Forest Ecology and Management*, Vol. 210, pp. 267-282, ISSN 0378-1127

Comiti, F., Andreoli, A., Lenzi, M.A. & Mao, L. (2006). Spatial density and characteristics of woody debris in five mountain rivers of the Dolomites (Italian Alps). *Geomorphology*, Vol.78, No.1-2, pp. 44-63, ISSN 0169-555X.

Densmore, N., Parminter, J. & Stevens, V. (2004). Corse woody debris: Inventory, decay modelling, and management implications in three biogeoclimatic zones. *BC Journal of Ecosystems and Management*, Vol.5, No.2, pp. 14-29, ISSN 1488-4666.

Duvall, M.D., Grigal, D.F. (1999). Effects of timber harvesting on coarse woody debris in red pine forests across the Great Lakes states, U.S.A. *Canadian Journal of Forest Research*, Vol.29, No.12, pp. 1926-1934, ISSN 0045-5067.

EC, (1993). *CORINE Land Cover technical guide, Report EUR 12585EN*. Office for Publications of the European Communities, Luxembourg.

FAO, (2004). *Global Forest Resources Assessment Update 2005: Terms and Definitions*. Rome: Working Papers 83/E, Forest Resources Assessment Programme.

Fridman, J., Walheim, M. (2000). Amount, structure, and dynamics of dead wood on managed forestland in Sweden. *Forest Ecology and Management*, Vol.131, pp. 23-36, ISSN 0378-1127.

Green, P., Peterken, G.F. (1997). Variation in the amount of dead wood in the woodlands of the Lower Wye Valley, UK in relation to the intensity of management. *Forest Ecology and Management*, Vol.98, No.3, pp. 229-238, ISSN 0378-1127.

Hagemann, U., Moroni, M.T. & Makeschin, F. (2009). Deadwood abundance in Labrador high-boreal black spruce forests. *Canadian Journal of Forest Research*, Vol.39, pp. 131-142, ISSN 0045-5067.

Hagan, J.M., Grove, S.L. (1999). Coarse Woody Debris. *Journal of Forestry*, Vol.97, No.1, pp. 6-11, ISSN 0022-1201.

Harmon, M.E., Sexton, J. (1996). *Guidelines for measurements of woody detritus in forest ecosystems*. Seattle: University of Washington, ISBN 952-458-128-0.

Harmon, M.E., Franklin, J.F., Swanson, F.J., Sollins, P., Gegory, S.W., Lattin, J.D., Anderson, N.H., Cline, S.P., Aumen, N.G., Sedell, J.R., Lienkaemper, G.W., Cromak, K. & Cummins, K.W. (1986). Ecology of coarse woody debris in temperate ecosystems. *Advanced Ecology Research* Vol.15, pp. 133-302, ISBN 0-12-013933-2.

Hegetschweiler, K.T., van Loon, N., Ryser, A., Rusterholz, H.P. & Baur, B. (2009). Effects of Fireplace Use on Forest Vegetation and Amount of Woody Debris in Suburban Forests in Northwestern Switzerland. *Environmental Management*, Vol.43, No.2, pp. 299-310, ISSN 0044-7447.

Hodge, S.J., Peterken, G.F. (1998). Deadwood in British forests: priorities and a strategy. *Forestry*, Vol.71, No.2, pp. 99-112, ISSN 0015-752X.

Holub S.M., Spears J.D.H. & Lajtha K. (2001). A reanalysis of nutrient dynamics in coniferous coarse woody debris. *Canadian Journal of Forest Research*, Vol.31, 1894-1902, ISSN 0045-5067.

Humphrey J.W., Sippola A.L., Lempérière G., Dodelin B., Alexander K.N.A. & Butler, J.E. (2004). Deadwood as an indicator of biodiversity in european forests: from theory to operational guidance. In: Marchetti, M. (eds) "Monitoring and Indicators of Forest Biodiversity in Europe – From Ideas to Operationality", *EFI Proceedings*, 51, pp. 193-206.

Hunter, M.L. (1990). *Wildlife, forests and forestry: principles of managing forests for biological diversity*. Prentice Hall, Englewood Cliffs, ISBN 13978013501432.

INFC, (2009). *I caratteri quantitativi 2005 - Parte 1*. In: Gasparini, P., De Natale, F., Di Cosmo, L., Gagliano, C., Salvadori, G., Tabacchi, G., Tosi, V. (eds). Inventario nazionale delle foreste e dei serbatoi forestali di carbonio, MiPAAF - Ispettorato Generale Corpo Forestale dello Stato, CRA-MPF, Trento, Italy.

Keller, M., Palace, M., Asner, G.P., Pereira, R. & Silva, J.N.M. (2004). Coarse woody debris in undisturbed and logged forests in the eastern Brazilian Amazon. *Global Change Biology*, Vol.10, No.5, pp. 784-795, ISSN 1354-1013.

Kirby, K.J., Reid, C.M., Thomas, R.C. & Goldsmith, F.B. (1998). Preliminary estimates of fallen dead wood and standing dead trees in managed and unmanaged forests in Britain. *Journal of Applied Ecology*, Vol.35, No.1, pp. 148-155, ISSN 0021-8901.

Kraigher, H., Jurc, D., Kalan, P., Kutnar, L., Levanic, T., Rupel, M. & Smolej, I. (2002). Beech coarse woody debris characteristics in two virgin forest reserves in southern Slovenia. *Zbornik gozdarstva in lesarstva*, 69, pp. 91-134, ISSN 0351-3114.

Krankina, O.N., Harmon, M.E. (1995). Dynamics of the dead wood carbon pool in Northwestern Russian boreal forests. *Water, Air and Soil Pollution*, Vol.82, No.1-2, pp. 227-238, ISSN 0049-6979.

Kueppers, L.M., Southon, J., Baer, P. & Harte, J. (2004). Dead wood biomass and turnover time, measured by radiocarbon, along a subalpine elevation gradient. *Oecologia*, Vol.141, No.4, pp. 641-651, ISSN 0029-8549.

Küffer, N., Senn-Irlet, B. (2005). Influence of forest management on the species richness and composition of wood-inhabiting basidiomycetes in Swiss forests. *Biodiversity and Conservation*, Vol.14, pp. 2419-2435, ISSN 0960-3115.

Laiho, R., Prescott, C.E. (1999). The contribution of coarse woody debris to carbon, nitrogen and phosphorus cycles in three Rocky Mountain coniferous forests. *Canadian Journal of Forest Research*, 29, pp. 1592-1603, ISSN 0045-5067.

Lombardi, F., Lasserre, B., Tognetti, R. & Marchetti, M. (2008). Deadwood in Relation to Stand Management and Forest Type in Central Apennines (Molise, Italy). *Ecosystems*, 11, pp. 882-894, ISSN 1432-9840.

Lombardi, F., Chirici, G., Marchetti, M., Tognetti, R., Lasserre, B., Corona, P., Barbati, A., Ferrari, B., Di Paolo, S., Giuliarelli, D., Mason, F., Iovino, F., Nicolaci, A., Bianchi, L., Maltoni, A. & Travaglini, D. (2010). Deadwood in forest stands close to old-growthness under Mediterranean conditions in the Italian Peninsula. *L'Italia Forestale e Montana*, Vol.65, No.5, pp. 481-504, ISSN 0021-2776.

Longo, L. (2003). "Habitat trees" and other actions for birds. *Proceedings of the International Symposium 29th-31st May 2003, Mantova (Italy)*, pp. 49-50.

Marage, D., Lemperiere, G. (2005). The management of snags: a comparison in managed and unmanaged ancient forests of the Southern french Alps. *Annals of Forest Science*, Vol.62, pp. 135-142, ISSN 1286-4560.

Marchetti, M., Lombardi, F. (2006). Analisi quali-quantitativa del legno morto in soprassuoli non gestiti: il caso di "Bosco Pennataro", Alto Molise. *L'Italia Forestale e Montana* Vol.61, No.4, pp. 275-301, ISSN 0021-2776.

Mason, F., Nardi, G. & Whitmore, D. (2005). Recherches sur la restauration des habitats du bois mort: l'exemple du LIFE "Bosco della Fontana" (Italie). In: Vallauri, D., André, J., Dodelin, B., Eynard-Machet, R. & Rambaud, D. (eds), *Proceedings of Bois mort et à cavités, une clé pour des forêts vivantes*, Éditions Tec & Doc, Paris, France, pp. 285-291.

Mehrani-Mylany, H., Hauk, E. (2004). Totholz - Auch hier deutliche zunahme. In:"Österreichische Waldinventur 2000/02 - Hauptergebnisse. *BFW-Praxisinformation*, No.3, pp. 21-23, ISSN 1815-3895.

Montes, F., Cañellas, I. & Montero, G. (2004). Characterisation of Coarse Woody Debris in Two Scots Pine Forests in Spain. In: Marchetti, M. (eds) "Monitoring and Indicators of Forest Biodiversity in Europe - From Ideas to Operationality", *EFI Proceedings*, Vol. 51, pp. 171-180.

Morelli, S., Paletto, A. & Tosi, V. (2006). Il legno morto dei boschi: prove di rilevamento campionario a fini inventariali. *Linea Ecologica*, Vol.38, No.3, pp. 51-57, ISSN 1721-9450.

Müller-Using, S., Bartsch, N. (2009). Deacy dynamic of coarse and fine woody debris of a beech (*Fagus sylvatica* L.) forest in Central Germany. *European Journal of Forest Research*, Vol.128, No.3, pp. 287-296, ISSN 1612- 4669.

Naesset, E. (1999). Relationship between relative wood density of Picea abies logs and simple classification systems of decayed coarse woody debris. *Scandinavian Journal of Forest Research*, Vol.14, No.5, pp. 454-461, ISSN 0282-7581.

Nordén, B., Ryberg, M., Götmark, F., & Olausson, B. (2004). Relative importance of coarse and fine woody debris for the diversity of wood-inhabiting fungi in temperate broadleaf forests. *Biological Conservation*, Vol.117, No.1, pp. 1-10, ISSN 0006-3207.

Paletto, A., Tosi, V. (2010). Deadwood density variation with decay class in seven tree species of the Italian Alps. *Scandinavian Journal of Forest Research*, Vol.25, No.2, pp. 164-173, ISSN 0282-7581.

Paolucci, P. (2003). Mammiferi e uccelli in un habitat forestale della pianura padana: il bosco della fontana. In: *Proceedings of the International Symposium* 29th-31st May 2003, Mantova (Italy), pp. 11-13.

Pignatti, G., De Natale, F., Gasparini, P. & Paletto, A. (2009). Il legno morto nei boschi italiani secondo l'Inventario Forestale Nazionale. *Forest@*, No.6, pp. 365-375, ISSN 1824-0119.

Radu, S. (2006). The ecological role of deadwood in natural forests. *Environmental Science and Engineering*, No.3, pp. 137-141, ISSN 1092-8758.

Raphael, M.G., White, M. (1984). Use of snag by cavity-nesting birds in the Sierra Nevada. *Wildlife Monographs*, Vol.86, pp. 1-66, ISSN 1938-5455.

Ravindranath, N.H., Ostwald, M. (2008). *Carbon Inventory Methods*. Handbook for Greenhouse Gas Inventory, Carbon Mitigation and Roundwood Production Projects. Springer, ISBN 1402065469, 9781402065460, New York, USA.

Rouvinen, S., Rautiainen, A. & Kouki, J. (2005). A relation between historical forest use and current dead woody material in a boreal protected old-growth forest in Finland. *Silva Fennica*, Vol.39, No.1, pp. 21-36, ISSN 1457-7356.

Sandström, F., Petersson, H., Kruys, N. & Ståhl, G. (2007). Biomass conversion factors (density and carbon concentration) by decay classes for dead wood of *Pinus sylvestris*, *Picea abies* and *Betula spp.* in boreal forests of Sweden. *Forest Ecology and Management*, Vol.243, No.1, pp. 19-27, ISSN 0378-1127.

Schlaghamersky, J. (2003). Saproxylic invertebrates of floodplains, a particularly endangered component of biodiversity. In: *Proceedings of the International Symposium* 29th-31st May 2003, Mantova (Italy): pp. 15-18.

Stokland, J.N., Tomter, S.M. & Söderberg, U. (2004). Development of dead wood indicators for biodiversity monitoring: experiences from Scandinavia. In: Marchetti, M. (eds) "Monitoring and Indicators of Forest Biodiversity in Europe – From Ideas to Operationality", *EFI Proceedings*, No.51, pp. 207-226.

Tobin, B., Black, K., McGurdy, L. & Nieuwenhuis, M. (2007). Estimates of decay rates of components of coarse woody debris in thinned Sitka spruce forests. *Forestry*, Vol.80, No.4, pp. 455-469, ISSN 0015-752X.

Van Wagner, C.E. (1968). The line intersect method in forest fuel sampling. *Forest Science*, Vol.14, No.1, pp. 20-26, ISSN 0015-749X.

Vallauri, D., André, J., & Blondel, J. (2003). Le bois mort, une lacune des forêt gérérs. *Revue Forestier Française*, No.2, pp. 99-112, ISSN 0035-2829.

Verkerk, P.J., Lindner, M., Zanchi, G. & Zudin, S. (2011). Assessing impacts of intensified biomass removal on deadwood in European forests. *Ecological Indicators*, Vol.11, No.1, pp. 27-35, ISSN 1470-160X.

Waddell, K.L. (2002). Sampling coarse woody debris for multiple attributes in extensive inventories. *Ecological Indicators*, Vol.1, No.3, pp. 139-153, ISSN 1470-160X.

Woldendorp, G., Keenan, R.J. & Ryan, M.F. (2002). Coarse woody debris in Australian forest ecosystems. Report for the National Greenhouse Strategy, Module 6.6 (Criteria and Indicators of Sustainable Forest Management), April 2002.

Wolynski, A. (2001). Significato della necromassa legnosa in bosco in un'ottica di gestione forestale sostenibile. *Sherwood*, 67, pp. 5-12, ISSN 1590-7805.

Zell, J., Kändler, G. & Hanewinkel, M. (2009). Predicting constant decay rates of coarse woody debris — A meta-analysis approach with a mixed model. *Ecological Modelling*, Vol.220, pp. 904-912, ISSN 0304-3800.

Close-to-Nature Forest Management: The Danish Approach to Sustainable Forestry

Jørgen Bo Larsen

Forest & Landscape, University of Copenhagen
Denmark

1. Introduction

How should tomorrow's forests look and which future climatic conditions should they prevail? What kind of goods, services and experiences should they be able to provide; what kind of functions should they be able to perform? These are some of the multifaceted questions forest management faces today.

Forestry policy objectives have grown into a broad range of benefit provisions, other than serving exclusively as the traditional timber suppliers. Today we thus address multiple-use forestry. Production of wood commodities and securing carbon storage is central, but does not necessarily rate above the creation of non timber forest products. Increasingly highly esteemed qualities, such as protecting landscape amenities and cultural heritage, nature conservation and environmental protection, as well as the entire chapter of recreational interests are considered. Consequently, economic and technical efficiency is still prioritized, but ecological and social parameters are progressively taken into account to ensure the multiple use.

For these reasons, silvicultural strategies are required to develop economically productive forests with a high potential for nature conservation, ecosystem protection, and social values. One promising management strategy is to incorporate structural qualities and functional features of natural forest ecosystems – "to follow and assist nature in her development" as already stated 230 years ago by the Danish forester von Warnstedt (Decree of 1781 regarding the management of the Royal forests). This approach can be summarised by the term "nature-based silviculture" or "close-to-nature management" (Gamborg & Larsen, 2003). In North America, on a more general forest management level, 'ecosystem management' and 'adaptive management' can be recognized as part of this trend (Franklin et al., 2002). The aim is to reform current practices so that they are still profitable, but more environmentally benign and even more sensitive to the complexities of nature conservation and the multiple, varying and steadily increasing demands of society by mimicking natural forest structures, their processes as well as their dynamics (Angelstam et al., 2004; Lindenmayer, et al., 2006; Hahn et al., 2007; Larsen et al. 2010).

The disturbances and regeneration processes in natural forest ecosystems, which cause structural heterogeneity at both large and small scale levels are linked to regional characteristics of climate, soil, and species compositions. These processes are being expressed as e.g. infrequent, large-scale storm and fire-driven disturbances in boreal ecosystems and as frequent, small-scale disturbances in Central-European forests. Hence models, describing the region-specific disturbance patterns, such as the forest cycle model, should be used in the development of applied silvicultural methods in such natural ecosystems (Hahn et al., 2005).

In central and western Europe the forest cycle models have been successfully applied to describe the temporal and spatial dynamics and cyclic preoccupation of a specific forest type in natural forest reserves (Leibundgut, 1984; Christensen et al., 2007; Larsen et al., 2010). Such models could serve as an adequate basis for close-to-nature forest management.

The use of natural disturbance regimes to guide human management (i.e thinning and cutting systems) must, however, be complemented with other measures to restore naturalness in forest management. Lindenmayer et al. (2006) emphasize the importance of maintaining aquatic ecosystem integrity for biodiversity protection in managed forests. Hence, maintaining and restoring natural hydrology in forests previously subjected to stand management operations (such as drainage) is important. Therefore promoting species and forest structures that reflect and emphasize the variation in hydrology is an integral part of close-to-nature management, thus contributing to habitat richness in forest landscapes.

One of the basic axioms of nature-based forestry is the mimicking of natural structures and processes in order to obtain a high degree of stability within the ecosystem and thus a high degree of flexibility. All of this necessary to opening up for possible future demands and needs from various players, such as landowners, interest groups and society in general. The logic of this assumption might be best illustrated by considering the contrary position: *without stability – which functions will we be able to sustain in the future.*

One of the major problems experienced with the classical forestry approaches, is the lack of ecological and structural stability and the limited flexibility, towards addressing various at present unknown future demands (who would 20 years ago have predicted the present focus on biodiversity in forest management?). An approach, which has alerted us of the necessity to search for better management systems aiming at increasing functionality as well as flexibility in forest ecosystems; both in relation to multiple uses. In other words: We seem to face a high potential for both ecological adaptability (resilience) and functional flexibility of forest ecosystems, when opening up for greater functional integration – a central aspect of close-to-nature management.

To illustrate the differences between the traditional plantation approach, the close-to-nature approach, and a strict nature protection approach, a "goal-fulfilment assessment" and comparison of the three different management approaches is shown in table 1.

Table 1 indicates that the plantation approach and the conservation approach both are rather narrow and inflexible in their goal-fulfilment, while the nature-based wood production approach is broad and flexible in its goal-fulfilment. The weaknesses of the plantation approach focussing on timber in short rotations and neglecting most natural, cultural and social values are clearly reflected in the table. Further, the plantation approach often leads to less robust forests stands. The conservation approach obviously performs strongly in all nature protection goals but is consequently not, or less able, to deliver on socio-economic goals thus scoring rather low in terms of flexibility to changing goals.

The maintained focus on production economy in combination with relative high scores in ecological as well as social values addressing the needs for stability, explains why the nature-based approach to sustainable forest management has been chosen in many countries. The integrative ability and flexibility of the nature-based approach to fulfil different management goals is a key feature of this management type. Because of this feature, it is possible to gradually adjust the course of management to address the ever-changing objectives and aspirations of society.

Management approach	Plantation (production) approach	Nature-based (integrative) approach	Nature protection (conservation) approach
Specific management goals	Focus on timber production and direct economic outcome	Flexible wood production, nature protection and recreation	Strict forest reserves following natural structures and processes
Production of timber	+++++	++++	+
Economic outcome, long term	+++	+++++	+
Economic outcome, short term	+++++	+++	+
Production of quality timber	++++	++++	+
Biodiversity protection	+	+++	+++++
Protection of wetlands	+	+++	+++++
Ecosystem integrity	+	++++	+++++
Aesthetic qualities	+	+++++	+++++
Landscape integration	++	++++	+++++
Historical and cultural values	+	++++	+++
Space for public recreation	++	++++	++
Place of quietness/meditation	+	+++	+++++
Hunting qualities	+++	++++	+
Robust and resilient forests	+	++++	+++++
Flexibility to changing goals	+	+++++	+

Table 1. Different management approaches and their respective fulfilment of different specific management goals. The scale from 1 to 5 plusses, and '+' = low goal fulfilment, whereas '+++++' = high goal fulfilment.

2. The history of nature-based forest management – In short

In Europe there have been attempts and local traditions to literally follow nature-near principle, to follow and steer the natural development in order to meet some more or less specific goals (Leibundgut, 1984; Schütz, 1990; Otto, 1993). However, the main trend in European forestry has followed the principles of organised forestry with a strong emphasis on clear-cutting, planting, thinning, homogenisation of structures, as well as rationalisation of working procedures. Organised forestry has had longstanding strong advantages in terms of overview, planning, standardisation, prediction and control.

While organised forestry has become the dominating concept in most parts of Europe, the more nature-inspired forestry approaches have been left to survive in the shade. Such concepts have not been given much attention, nor has much research been carried out to highlight possible advantages of this branch of silviculture. The ideal "to follow and assist nature in her development" has often been cited – but in reality rarely been followed in practice. Until recently, most attempts to apply nature-based forestry have been mainly exceptions from the rule. They have been carried out under special conditions and have been conducted by individuals mainly driven by conviction. A belief which has led to the assumption that nature-base forest management could turn out to be a more promising approach than traditional plantation forestry. People practising nature-based forestry have thus in the recent past often been given the image of being some kind of "religious freak" (Heyder, 1986).

Close-to-nature forestry is unquestionably focused around the idea of selection forest. The single tree and group selection system represent a clear contrast to the even-aged forests of organised forestry. Many foresters have tried to develop such uneven-aged mixed forests and have searched for appropriate methods to evaluate management successes, in order to compare them objectively to even-aged systems. The French forester Adolphe Gurnaud (1825-1898) once succeeded with the French *Méthode de contrôle*. His method based on regular inventories of forests parameters, especially diameter distribution and increment. Although not successful in implementation of his ideas, Henri Biolley (1858-1939) later succeeded in managing the community forest of Couvet with this "modern" selection system (Biolley, 1920).

Another important source of nature-based forestry started around the ideas of Karl Gayer (1822-1907), a silviculture professor in Munich. At that time, organised forestry with clear-cut systems and introduction of conifers had already expanded over large forest areas. Consequently, following this process, soil degradation, fungi and insect outbreaks, as well as frequent windbreaks had been observed in those areas. As a reaction, Gayer then developed his idea of mixed forests, which were about to be achieved merely through natural regeneration (Gayer, 1886), often in combination with the irregular shelterwood system. Using irregular regeneration over a longer time-span would thereby enable various different species to establish and thereby creating mixed forest structures.

His ideas were further developed in Switzerland. At that time Swiss forests suffered severely from torrents, landslides and windbreaks, as a result of spruce monocultures and clear-cut management systems. Arnold Engler (1858-1923) succeeded in gradual change of the Swiss forestry paradigm, which was untied from the regeneration scheme of organised forestry.

Today, variations of the Swiss irregular shelterwood systems are the most widely applied nature-based silvicultural systems all over Central and Eastern Europe. This is mainly thanks to the great flexibility of the system, which is based on the principles of adapting the felling temporally and spatially to the regeneration ecology of various tree species. Apart from the selection system, the irregular shelterwood system for nature-based forestry and the "free-style silvicultural technique" are significant as well; especially when it comes to managing degraded forests or transforming uniform and even-aged forests into mixed uneven-aged forests.

The third "wave" of nature-based forestry has developed arund 1920 in northern Germany when Alfred Möller published the book "Der Dauerwaldgedanke" (Möller, 1922). His paradigm of a continuation forest differs essentially from other nature-based concepts. Möller's approach is based on an organismic and holistic conception of forests and it follows stricter felling rules. His ideas had been shaped in forests where careful, continuous forest cover forest management had been applied for many years (Bärenthoren in eastern Germany). Möller carries out different inventories and publishes his results in favour of continuous forest cover management (German: Dauerwald), which, according to his conviction, offer improved forest sites, abundant regeneration as well as increased wood production.

Möller's forest approach was welcomed with great sympathy during the first years after his book had been published. 'The Dauerwald concept' was embraced with great enthusiasm all over Germany. When Möller died soon after publishing his book and his ideas proved unable to deliver the hoped success in the field, his approach became increasingly questioned and in the end even doubts about his scientific credibility ended this chapter of nature-based forestry in Germany during the 1930's.

With the foundation of a working group for close-to-nature forestry (Ger. Arbeitsgemein-schaft naturgemäße Waldwirtschaft - ANW), in 1950 yet another force steps onto the forest management scene. The ANW was rooted in the Dauerwald movement, and the groups ideas based on a set of principles rather than a management system. The group's members are mainly practising foresters and forest owners. Decisions on how to manage forests and strategies are empirical and often intuition based.

The call for for expanding the ANW-movement outside Germany resulted in the foundation of Pro Silva Europe in Slovenia in 1989. Pro Silva advocates close-to-nature forest management based on natural processes. Most European countries (at present 24) have joined and established national, independent Pro Silva sub-organisations. Their common ground on the national level is to develop and promote the principles of sustainability. These principles are considered to allow for the full development of the forests ecological and social roles, while a simultaneous economic production of high quality forest products can take place - all by mimicking natural processes. Members are forest owners, foresters, students and others who wish to practice and learn more about nature-near forestry.

2.1 The toolbox of classical and nature-based forestry

Basically classical plantation silviculture and close-to-nature approaches make use of the same toolbox in managing forests. However, the importance of single tools differs between the two concepts. Table 2 illustrates how different silviculture tools can be applied and combined under different management approaches. The plantation approach is displayed in two versions: traditional and modified (to achieve a higher degree of sustainability). Accordingly, the nature-based approach is displayed in a more economic, and a conservation focussed version.

Table 2 shows how management approaches determine what tools might be appropriate and most widely used. It further shows that although it is meaningful to differentiate between the various management approaches, it is neither possible, nor is it meaningful to draw a strict watershed line between those definition categories. Naturally transgression corridors occur. For each strategy however, it is possible to provide a set of relevant silviculture tools. Depending on management styles and aims within plantation, respectively nature-based management, the relative importance of the different tools can be adjusted. Each forest owner and each policy maker must critically choose his or her favourite tools for the situation and objectives which are being focussed upon.

Silviculture tool / Anticipated stand structure	Traditional plantation (production) approach	Modified (sustainable) plantation approach	Nature-based economic production approach	Nature-based nature conservation approach
Clear cutting at rotation age	+++++	++++	++	+
Single tree/group cutting	+	++	+++++	+++++
Planting or sowing	+++++	++++	++	+
Natural regeneration	+	++	++++	+++++
Use of soil preparation	+++++	++++	++	+
No soil preparation	+	++	++++	+++++
Use of pesticides	+++++	++	+	+
Ban of pesticides	+	++++	+++++	+++++
Use of exotic species	+++++	++++	+++	+
Use of native species	+	++	+++	+++++
Stand management	+++++	++++	++	+
Single tree management	+	++	++++	+++++
Harvest when ripe	+++++	++++	++++	+
Preserving old trees	+	++	++	+++++
Wood salvage	+++++	++++	++++	+
Leaving dead wood	+	++	++	+++++
Draining for production	+++++	++++	+++	+
Maintain wet habitats	+	++	+++	+++++
Monoculture	+++++	++++	++	+
Species mixtures	+	++	++++	+++++
Even-aged stands	+++++	++++	++	+
Uneven-aged stands	+	++	++++	+++++

Table 2. Examples of silviculture tools and anticipated stand structure and their relative importance in nature-based as well as classic (plantation) forest management:
+++++ greatly used, ++++ frequently used,   +++ regularly used,   ++ rarely used;
+ hardly ever used

Nature-based approaches in general refrain from larger clear cuts, but in specific cases - often in order to promote light demanding (pioneer) species - clear-cuts can be applied. Nature-based management relies heavily on natural regeneration but includes planting or direct seeding if natural regeneration is insufficient and/or if desired species are missing (enrichment planting). Nature-based approaches often make use of single tree selection based on target-diameter cutting, which should not be misinterpreted as "high grading" known from overexploitation of natural stands. Thus we here focus on a system to provide a sustained yield by making thinning among the various age classes in order to ensure their desired proportions and to maintain a suitable mixture of species. It should further be stressed that the heterogeneity of the stands is not just an end in itself, but rather a way of allocating species to various soil conditions and creating good forest floor conditions for natural regeneration.

The toolbox concept implies the refrain from any specific (religious) interpretation of what nature-based forest management is or should be - rather, the toolbox should be open to anyone finding the tools appropriate for any use he or she might wish for. The tools from this nature-based toolbox can be used for nature-protection, for wood-production or to develop new types of urban forest (Larsen and Nielsen, 2011).

However, the main prerequisite for defining an approach "nature-near" or "close-to-nature" should be *that the practices are founded in, or inspired by, the structures and processes that occur in natural forests of a specific (reference) region*. This principle can be used to achieve all kinds of different management goals and objectives, including timber production, nature protection and social values.

3. Close-to-nature forest management in Denmark

The forests in Denmark amount to a total area of 570.800 ha, equivalent to some 13 percent of the total land cover. Originally, most of the land has been forested, but after centuries of uncontrolled logging and deforestation for agriculture, forest areas begun to decline drastically and consequently collapsed to a mere 2 to 3 percent around the 1820's. Since then the forest area is increasing due to large forestation efforts from 1860 and onwards and expected to reach around 20 % within this century.

Originally the Danish forest consisted mainly of deciduous trees - especially beech and oak. Over the past 200 years of forest management - including the large forest plantings in Central and West Jutland – the species distribution changed radically. Today, more than 50 % of forested areas are covered with non native conifers such as Norway and Sitka spruces, Douglas fir, as well as different Abies-species. Deciduous forests cover not more than 44 %, with beech and oak as the most common species, and ash, sycamore maple, Norway maple, birch, alder, wild cherry and lime as minor species.

As a general trend, forestry in Denmark has followed the overall European development when focussing on timber production in mostly plantation like structures. As a result, highly productive forests have been promoted, a process, which simultaneously created the matrix for increasingly intense conflicts with nature protection interests. First and foremost, the stability of the forests suffered through the development of even-aged monocultures. During the last 40 years, in 4 storms (1967, 1981, 1999, and 2005) a total of 15 million m³ were blown down, whereas "only" 1 million m³ fell down during the first 60 years of the past century. Hence, a major reason for the increasing impact of storms in Danish forests is the increasing use of storm sensitive conifer species.

In order to realize sustainable forestry at the management unit level (to achieve a proper balance between economic, ecological and social functions), a set of overall aims and operational guidelines has been developed in a stakeholder driven process during 2001. The National Forest Programme (Skov- og Naturstyrelsen, 2002) now consequently prescribes that Danish public forests should be managed in accordance with close-to-nature principles. The essence in these close-to-nature principles can be summarized as follows: Increase the stability and prepare the forests for an unknown future of changing climate, changing values and a variety of goals.

This close-to-nature approach is in particular focussed on:

1. Creating optimal conditions for natural regeneration by maintaining the permanent forest climate by refraining from clear-felling.
2. Stability improvement and risk diversification (resilience) through the creation of uneven aged mixed forest stands of site-adapted tree species.
3. Active stand improvement through frequent and weak thinning.

4.   Protection of natural equilibriums among forest organisms, including pests, with the aim of promoting biodiversity and avoid the use of pesticides.

The close-to-nature forest management, combined with an increased use of climate robust deciduous and coniferous species and the reduction of climate change intolerant conifers (i.e. Norway spruce and Sitka spruce), are here identified as the overarching principles to secure sustainability, safeguard stability, and prevent the negative effects of climate change. Consequently, The Forest Act from 2004 supports the change from classical mono-species and even-aged management of stands into close-to-nature management characterised by more single tree and group management, incorporating and supporting natural regeneration and structural differentiation.

This decision to transform "classical" age-class forests (plantation forestry) towards nature-based forest stand structures implied no less than a paradigm shift in the management of state owned forests. Realizing that the complex character of these near-natural forest structures and dynamics require integrative and flexible management frameworks, as well as tools, a two step process was established: Firstly, the need for defining and describing long term goals for nature-near stand structure and dynamics was recognized and taken into the picture (where are we going?). Secondly, methods for transformation from plantation to nature-near structures were specified (how do we get there?).

3.1 The long-term goals – Creating Forest Development Types (FDT)

The concept "Forest Development Type" (FDT) was considered as an adequate framework for advancing and describing long-term goals for stand structures and dynamics in stands subjected to close-to-nature management (Larsen and Nielsen, 2007). An FDT describes the direction for forest development on a given locality (climate and soil conditions) in order to accomplish specific long-term aims of functionality (ecological-protective, economical-productive, and social-/cultural functions). It is based upon an analysis of the silvicultural possibilities on a given site in combination with the aspirations of future forest functions. It will serve as a guide for future silvicultural activities in order to "channel" the actual forest stand into the desired direction. Such a common understanding and agreement upon the desired development is crucial, since the conversion from age-class to nature-based stand structures is a continuous process.

In Denmark, a participatory process lead and described by Larsen and Nielsen (2007) resulted in the creation of 19 FDT's, which can be grouped into 9 broadleaved dominated, 6 conifer dominated, and an additional 4 "historic" types (Table 3). Whereas all "nature-based" FDT encompass a balance between productive, protective and recreational/social functions, the other four "historical" types mainly serve to protect recreational, natural and cultural functions. Especially the historical Forest Pasture (FDT No. 92) and Forest Meadow (FDT No. 93) can be actively used to create habitat diversity and experiential richness in forest landscapes.

Each FDT is described as follows (See also Figure 2, describing FDT No. 12 "Beech with ash and sycamore"):

* Name: The name encompasses the dominating and co-dominating species. The first digit in the FDT-number indicates the main species (1 = beech, 2 = oak, 3 = ash, 4 = birch, 5 = spruce, 6 = Douglas fir, 7 = true fir, 8 = pine, and 9 indicating a "historic" FDT). The second digit is numbered at random.
* Structure: A description of how the forest structure could appear when fully developed. This description is supplied with a profile diagram depicting a 120 m transect of the

anticipated forest structure at "maturity" (In Figure 1 profile diagrams of all 19 FDT´s are displayed and in Figure 4 the profile diagrams of four FDT´s are with different forest-edge types shown: No. 11-Beech, No. 21-Oak with ash and hornbeam, No. 71-Silver fir and beech, and No. 92-Forest pasture).

- Species distribution: The long-term distribution of species and their relative importance.
- Dynamics: The regeneration dynamics described in relation to the expected succession and spatial patterns (species, size).
- Functionality: Indication of the forest functionality (economic-production, ecologic-protection, and social/cultural functions).
- Occurrence: Suggested application in relation to climate and soil. For this purpose the country is divided into 4 sub-regions each with their typical climatic characteristics. Further, the application of the specific FDT in terms of soil conditions is stated in relation to nutrient and water supply.

Broadleaved dominated:	Conifer dominated:
11 Beech	51 Spruce with beech and sycamore
12 Beech with ash and sycamore	52 Sitka spruce with pine and broadleaves
13 Beech with Douglas fir and larch	61 Douglas fir, Norway spruce and beech
14 Beech with spruce	71 Silver fir and beech
21 Oak with ash and hornbeam	81 Scots pine with birch and Norway spruce
22 Oak with lime and beech	82 Mountain pine
23 Oak with Scots pine and larch	**"Historic" forest types:**
31 Ash with alder	91 Coppice forest
41 Birch with Scots pine and spruce	92 Forest pasture
	93 Forest meadow
	94 Unmanaged forest

Table 3. The 19 Danish Forest Development Types.

Matching forest development types to site

While different forest development types possess different site requirements it is possible to address and utilise potential variation in site conditions by matching FDT to site. This requires a thorough site survey, in which analyzing the basic growth conditions such as geology and soil types, nutrient and water supply, as well as specific site factors (such as compact layers and insufficient drainage) are taken into account. A hydrological status analysis on site is necessary, and it should include a survey of existing drainage systems, in combination with a plan of the historic landscape with former wet-lands, prior to any draining process. This hydrological analysis will provide an important tool and inspiration for delineating the landscape into ecological functional units. The site classification map works correspondingly as a frame for applying FDT to the site, thus facilitating the creation of forested landscape where site adapted forest and nature types reflect and emphasize variations within landscape.

Further, different FDTs possess different combinations of goal fulfilment - some are more production oriented, some more oriented towards nature/biodiversity protection, while others focus on enhancing landscape and recreational values. This variation in goal achievement can correspondingly be used to select FDTs - all according to specific functional requirements defined by the forest owner and - in case of public forest - by society/interest groups.

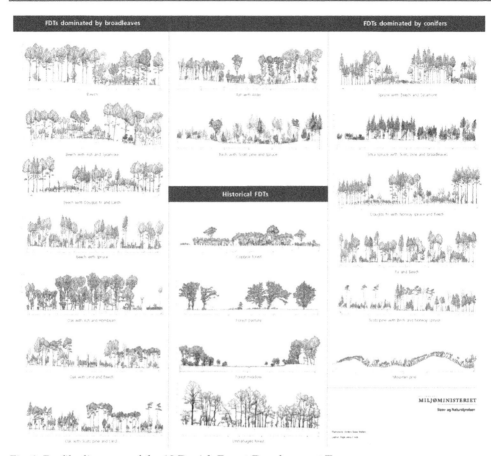

Fig. 1. Profile diagrams of the 19 Danish Forest Development Types.

At present the Nature Agency, responsible for the management of the Danish state forests, is laying out a grid system of forest development types on all public forests. This grid system will provide the local forest manager with information about the long-term goals he should aim at in each and every part of "his" forest. The managers job as local silviculturist will consequently be to observe the natural development and only then, after having conducted his observational research, to start making adjustments (cutting, planting, weeding, fencing, soil scarification etc.) in case the stand is due for short-term economic intervention (commercial thinning) and/or the actual development compromises the long term goal, as described in the attributed forest development type.

As mentioned above, the process of marking out FDTs on a management unit level is at present ongoing in Denmark. To illustrate this process, as well as the outcome, an example will be shown below. This example inspects the FDT-plan for the eastern part of Vestskoven as proposed by a group of students attending the international master course in Urban Woodland Design and Management (plan described in detail in Larsen & Nielsen, 2011).

Vestskoven was established in the 1960's west of Copenhagen to create a large recreational forest that could separate and structure the intense and rapid urban sprawl, and provide for

Forest Development Type 12: Beech with ash and sycamore

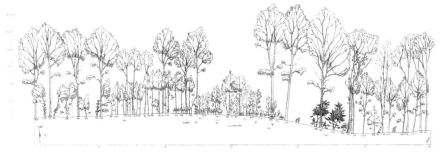

Structure: Species rich, well structured forest with beech as dominating element mixed with ash and cherry and in south-eastern Denmark additionally with hornbeam and lime. The in-mixed species occur mainly in groups. The horizontal structures arise between groups of varying size and age. Where the light demanding species such as ash, sycamore and cherry dominate, vertical structures occur periodically with shade trees (beech, hornbeam, elm, and others) in sub-canopy strata.

Species: Beech. 40 – 60 %, ash and sycamore: 30 – 50 %, cherry, hornbeam, oak, lime, and others up to 20 %

Dynamics: Beech regenerates mainly in groups and smaller stands. Ash and sycamore as gap specialists regenerate in openings later followed by beech. Hornbeam belongs to the sub-canopy stratum and regenerates under shade, whereas the pioneer species (cherry and oak) only regenerate after lager openings and/or in relation to forest edges.

Functions: Productive: The forest development type has a high potential for production of hardwood in larger dimensions and of good quality.

Protective: In most parts of the country the beech dominated forest represents the potential natural vegetation; consequently, many indigenous species are connected to this forest development type. It has a great potential for conserving biodiversity connected to the NATURA 2000 habitat type 9139 and 9150.

Recreational: Through ist mixture of (indigenous) species in combination with pronounced variation in size the forest development type gives a multitude of recreational experiences and intimacy.

Occurrence: The forest development type belongs on protected sties in the eastern and northern parts of Denmark on rich, well drained soils with good water supply as illustrated below.

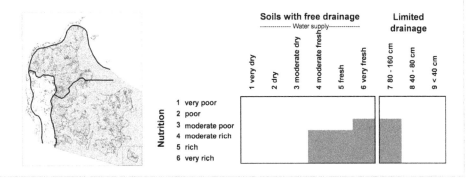

Fig. 2. Description and illustration of Forest development Type 12: Beech (*Fagus sylvatica*) with ash (*Fraxinus excelsior*) and sycamore (*Acer pseudoplatanus*).

important recreational qualities for the 300.000 new citizens in the western parts of Copenhagen. Fields were planted successively as they were purchased; little consideration was given to the overall composition and interlock zones between stands, or those parts dividing forested from open areas. The fields were planted according to traditional manuals with monoculture stands or simple species mixtures, using the species that were available at nurseries. The forest thus consists of small stands with abrupt species transitions and edges, all together lacking valuable interlock zones between the forested and more open areas. Today the area functions as a traditional Danish timber production forest with some large open spaces for recreation sprinkled onto it (Figure 3).

Fig. 3. Photo (from east towards west) of the eastern part of Vestskoven, showing the fragmented composition of uniform blocks of geometrically shaped stands and open spaces.

The above description demonstrates that Vestskoven incorporates most of the potentials, but even many problems, which urban woodlands inherited from the commercial forest management tradition with its uniform stand structures and its fragmented blocks of geometrically formed stands and open areas. The absence of smaller openings and glades, and the lack of valuable wetlands thus mould a fragmented, disconnected forest landscape.

Since Vestskoven is a public forest it will be managed according to close-to-nature principles and it is currently in the process of being charted into the FDT grid. Figure 4 presents a conversion/restoration plan where four Forest Development Types (FDT´s) have been laid out in respect of existing values in the young plantations and adjacent plains. The four selected FDT´s (FDT 11, Beech; FDT 71, Silver fir with beech and spruce; FDT 21, Oak with ash and hornbeam; FDT 92, Grazing forest), each with distinct experiential and ecological characteristics, unify the many small stands within larger units. The variety of size in open areas is increased by adding small, intimate glades in the forested parts. Some of the open areas have been linked to add further spatial variation and to increase coherence.

Parts of the forested, as well as the open areas have been converted into grazing forest through heavy thinning and some additional planting of trees. The borders between forested parts and open areas have been re-shaped organically by cutting out some of the existing stands, and instead giving room for edge species in those corridors. Thereby important interlock zones are being shaped between the denser forested and the more open areas, allowing for more diverse and complex edge structures. Ponds have been restored at emerging wetlands to render valuable landscape attractions, both in regard to landscape interpretation by visitors, as well as in regard to biodiversity in general. This landscape re-shaping takes place in the vicinity of small glades, at forested edges and in larger plains.

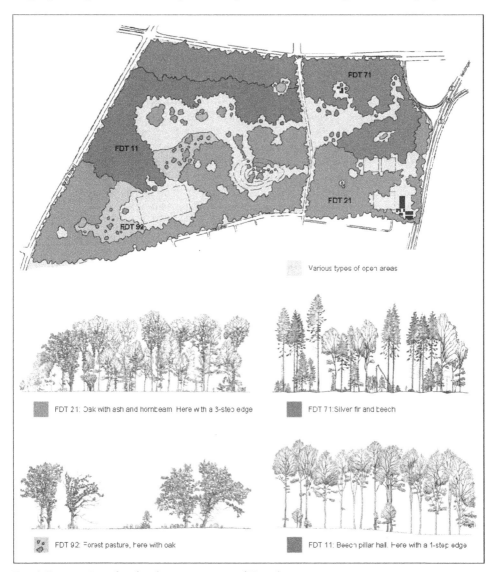

Fig. 4. Restoration plan for the eastern part of Vestskoven:

This plan was developed by a group of students attending the international master course in Urban Woodland Design and Management (Larsen & Nielsen, 2011). The chart, in combination with the profile diagrams of the four FDT´s, including examples of different edge-types, gives an instant impression of the anticipated urban forest landscape goals.

Furthermore, it provides an outline for appropriate developments in different parts of the forest. Such a developed and augmented plan, in combination with an FDT-map and profile diagrams of the different forest development types applied, can be used in multiple participatory planning processes.

3.2 Conversion principles and methods

Having defined the long-term goal at each part of the forest, the practitioners' principal task is to "guide" the forest from the current structure toward the targeted FDT. To help the local manager in this new endeavour, a number of conversion models haven been developed through a participatory process with local practitioners, forest workers and entrepreneurs. The primary purpose of this process is to come up with ideas as to how the conversion of a number of typical output models toward the desired forest type of development can take place. Since the conversion of uniform stands of spruce and beech are the main challenges in the transition to close-to-nature forest management in Denmark, the emphasis is on models for these species. Therefore, it is important to emphasise that these models are intended only to be used as inspiration, and they will always have to be adapted to any local situation, as well as to the concrete economic and technical possibilities. Especially the pace, at which the conversion is to be preformed, must be thoroughly analysed in regard to any economic aspects, paying special attention not to compromise expectation values for wood production in the transition phase. Therefore, in most cases, the full transformation to nature-near structures might take up to one or two tree generations.

Deciding on conversion strategy and tools there are two fundamentals, which must be kept in mind: Firstly, stand stability must be ensured and natural regeneration conditions must be improved. Thus creating various options and "freedom," timely to initiate rejuvenation (including bringing in new species), if required. Secondly, it is essential to initiate these elements at the ecologically and economically right moment in time. Thus, we speak of 1) a preparation phase, where the forest is stabilised and prepared for regeneration - mainly through selective thinning operations, and 2) a transformation phase, characterized by passive or active initiated regeneration, respectively by introduction of new species (and the procedure of ensuring their development). The preparation phase is usually associated with income (or at least cost neutrally implemented); whereas the transformation phase often entails costs (investments). Although, according to the principles of biological rationalization (a central economic aspect of close-to-nature-management), these costs could be kept on minimized levels by letting nature itself do as much of the "work" as possible.

If the forest development type prescribes species which are not present, or their genetic constitution (provenance) is not acceptable, additional seeding or planting (enrichment) in groups (typically, beech, ash, maple, birch, bird cherry, fir, larch, Douglas fir, etc.) is foreseen. These groups can later on contribute to a more widespread distribution of the species (done through seed dispersal). In order to allow rejuvenation of stable, but often frost-sensitive species (beech, firs including Douglas fir, etc.), a continued forest climate is regarded vital. Under such circumstances a stable forest canopy is paramount; especially in critically exposed, storm sensitive spruce monocultures. If a more complete conversion to new main tree species is aimed at already in the first generation, an extra widespread planting or seeding is envisaged, but often at higher cost. However, the close-to-nature approach is in general more inclined to exploit cheap regeneration methods, thereby accepting a longer conversion phase.

Generally, we distinguish between passive and active conversion strategies. The passive strategies are primarily based on existing vegetation, in order to convert as economically efficiently as possible. This implies mostly long conversion periods (up to several tree generations). The active approach is used where stability does not allow a slow (pending) conversion and/or there are other motives (ecological, aesthetical, and recreational) that advocate for a fast conversion.

Passive strategies

The purpose of the passive strategies is to implement as low-cost rejuvenation as possible, while maintaining optimum production in the upper canopy and the area as a whole.

Exhibit stands a high degree of stability; a passive strategy can be used that largely exploits the stand productive potential for transition to target diameter cutting without losing the possibility of a conversion. Transition phase can likewise extend over a long period of time, utilizing the system's own forces (natural regeneration), supplemented with scattered introduction of "new" species, if needed in the emerging gaps. Under such conditions, there are usually no major conflicts between the long-term objectives and the operating economy of the conversion phase. Gap size, and thus the potential light radiation, plays a crucial role in the choice of implanted species where they do not appear spontaneously. Thus, light demanding species such as larch, Douglas fir, oak, birch etc. require larger gaps (above 0.4 ha), while in the smaller gaps (0.1 - 0.2 ha), more shade tolerant species such as beech, maple and fir will be suitable.

Fig. 5. Passive approach; Spontaneous regeneration of fir, spruce, birch, larch and Mountain ash in wind-throw gaps in a Norway spruce stand (group regeneration), Klosterheden Statsskovdistrikt. Photo: J.B. Larsen

Active strategies

The active conversion approach is used under conditions, where lack of stability does not allow a passive conversion. Active strategies are used in unstable stands primarily of spruce. In potentially unstable stands which have not yet reached a height that makes them storm exposed (below approx. 14 m), it is important to conduct an active thinning to promote stability and structural variation. This can happen partly through an early shelterwood formation or by liberating a number of future trees, thereby creating stable single trees (anchor trees). Important is that the thinning is conducted "from above" (removing dominating and co-dominating trees) thereby promoting variation in tree size (diameter, height) and a more heterogeneous stand structure. Group felling, in combination with early introduction of regeneration are also examples of active strategies. It is common to these approaches that a portion of the potential production in the stand will be sacrificed to safeguard the success of regeneration. In some cases the only economically realistic approach for regeneration/conversion of unstable spruce stands will be a clear cut; a measure, which also can be considered as an active strategy. In situations, when clear-cutting is the only way to regenerate the stand, frost hardy pioneer species such as Scots pine, oak, larch and birch will be introduced by planting/sowing to supplement, to improve the frequent natural regeneration of spruce and birch, thereby increasing future silvicultural options and thereby successively moving towards the planned long term goal – the FDT.

Fig. 6. Active approach; 9-year old beech planted under a canopy of Norway spruce (shelter-wood regeneration). Klosterheden Statsskovdistrikt. Photo: J.B. Larsen.

The choice of conversion strategy depend on the starting point including the potential stability of the concrete stand, the objective defined by the FDT, and the time available for the conversion according to the economic perspective of net-present values of anticipated functions, together with the conversion costs. In total 10 different conversion and regeneration models have been developed for converting monocultures of beech, spruce and oak into nature-near structures. In Figure 7 such two models are displayed by means of

Left: Passive approach showing the conversion of a 54-year old even-aged beech stand to FDT 12 – Beech with ash and sycamore maple. The so-called "qualitative group cutting" is applied. Thinning is preformed by cutting trees according to their quality disregarding an even distribution of the remaining trees. This will create openings in the closed beech stand, where ash and maple is introduced. The regeneration is completed by natural regeneration slowly creating a group-wise structure of beech, ash and maple.

Right: Active approach showing the conversion of a 24-year old Norway spruce plantation to FDT 61 – Douglas fir, Norway spruce and beech. The thinning regime aims at creating variation in the overall thinning density in the area. It is done to open up for creation gaps, to be filled with Douglas fir and beech. The rest of the area is regenerated naturally with spruce and birch.

Fig. 7. Conversion models displayed with profile diagrams.

profile diagrams, depicting a possible development from a uniform plantation like structure towards the decided nature-near forest development type.

4. Conclusion

The management of forests "closer to nature" has increased significantly in recent decades, simultaneously accompanied by ever more reliable and refined models, promoting its efficient implementation. The basic idea is to reach a better balance between productive, protective and social functions. Other important goals are to increase economic competitiveness by cost reduction and increase robustness to climate change.

In Denmark, the Nature Agency started to manage all public forest according to close-to-nature principles in 2005. To facilitate the transition from classical even-aged plantation forestry to close-to-nature silviculture a total of 19 Forest Development Types (FDTs) and different conversion models have been developed in a participatory process with forest practitioners, scientists, forest workers, contractors and other stakeholders.

Now, almost 10 years after the political initiation, and 6 years after the state forest once started to be managed according to close-to-nature principles, the picture is multifaceted: The conversion process in the state forests is continuing with special focus on developing nature rich recreational forest landscapes, by means of the FDT planning scheme. A massive effort to restore natural hydrology is one of the most significant ingredients in the process; as well as the integration of permanent open spaces in the forest (forest meadows – FDT 93), the introduction of grazing animals (forest pasture – FDT 92), and the delineation of larger reserves (unmanaged forest - FDT 94). Furthermore, different methods and models for converting spruce plantations have been used. Still, it seems too early to draw any final conclusions in regard to his last aspect. The lack of funding for a scientific follow-up is a potentially jeopardising aspect.

Many forests belonging to municipalities have also changed management strategies fundamentally and they now apply the close-to-nature silviculture guidelines. Especially the FDT planning tool-box has proven highly effective to generate discussion platforms to define goals and ways of forest management among various stakeholders in urban forests.

The private forest sector is still rather reluctant in applying close-to-nature management. Some forest owners are doing it with great enthusiasm, while a majority still sticks to the classical age-class plantation system. However, the running debate about the pros and cons has had its effect on the size of clear-cuts and the use of natural regeneration.

We are learning by doing: Some of the pending issues are: How much reduction in professional input/contribution is possible without loosing the advantages of close-to-nature management? To what extend is it possible to educate private forest contractors to apply close-to-nature silviculture with their big machines? Is it possible to create the same high wood quality in un-even aged forest systems as in plantation like structures – and to what costs? How can the close-to-nature managed forest cope with the increased need for bio-energy production?

5. Acknowledgments

The author wants to thank Dagmar Nordberg for valuable contributions and Alan & Jane Newbury for proof reading.

6. References

Angelstam, P.; Boutin, S.; Schmiegelow, F.; Villard, M.-A.; Drapeau, P.; Host, G.; Innes, J.; Isachenko, G.; Kuuluvainen, T.; Mönkkönen, M.; Niemelä, J.; Niemi, G.; Roberge, J.-M.; Spence, J. & Stone, D. (2004). Targets for boreal forest biodiversity conservation – a rationale for adaptive management. *Ecoogical Bulletins.* Vol. 51: pp 487–509.

Biolley, H. (1920). L'aménagement des forets par la méthode expérimentale et spécialement la méthode du contrôle. *Beiheft Schweizerischer Forstverein,* Vol. 66, 1980, pp 51-134.

Christensen, M.; Emborg, J. & Nielsen, A.B., (2007). The forest cycle of Suserup Skov - revisited and revised. *Ecological Bulletins,* Vol. 52, pp 33-42.

Franklin, J.F.; Spies, T.A.; Van Pelt, R.; Carey, A.B.; Thornburgh, D.A.; Berg, D.R.; Lindenmayer, D.B.; Harmon, M.E.; Keeton, W.S.; Shaw D.C.; Bible, K. & Chen, J. (2002). Disturbances and structural development of natural forest ecosystems with silvicultural implications, using Douglas-fir forests as an example. *Forest Ecology and Management,* Vol. 155; pp 399-423.

Gamborg, C. & Larsen, J.B. (2003). 'Back to nature' a sustainable future for forestry? *Forest Ecology and Management,* Vol. 179, pp. 559-571.

Gayer, K. (1886). *Der gemischte Wald.* Verlag von Paul Parey, Berlin, 168 pp.

Hahn, K.; Emborg, J.; Larsen J.B. & Madsen, P. (2005). Forest rehabilitation in Denmark using nature-based forestry. In Stanturf J.A., and Madsen P. (eds.): *Restoration of boreal and temperate forests.* CRC Press, 299-317.

Hahn, K., Emborg J.; Vesterdal L.; Christensen S.; Bradshaw R.H.W.; Raulund-Rasmussen K. & Larsen J.B. (2007): Natural forest stand dynamics in time and space - synthesis of research in Suserup Skov, Denmark and perspectives for forest mangement. *Ecological Bulletins.* Vol. 52, pp. 183-194.

Heyder, J.C. (1986). *Waldbau im Wandel.* J.D. Sauerländer's Verlag, Frankfurt am Main.

Larsen, J.B. & Nielsen, A.B. (2007): Nature-based forest management – where are we going? – Elaboration forest development types in and with practice. *Forest Ecology and management,* 238, 107-117.

Larsen, J.B.; Hahn, K. & Emborg J. (2010). Forest reserve studies as inspiration for sustainable forest management – Lesson learned from Suserup Skov in Denmark. *Forstarchiv,* Vol. 81, pp 28-33.

Larsen, J.B. & Nielsen, A.B. (2011): Urban forest landscape restoration - Applying Forest Development Types in design and planning. In: *Forest Landscape Restoration: Integrating Natural and Social Sciences.* Springer Publishing Company. Accepted

Leibundgut, H., 1984: *Die Waldpflege.* Verlag Paul Haupt Bern, Stuttgart, 216 pp.

Lindenmayer, D.B.; Franklin, J.F.; & Fischer, J. (2006). Conserving forest biodiversity: A checklist for forest managers. *Biological Conservation,* Vol. 129, pp 511-518.

Möller, A., 1922: *Der Dauerwaldgedanke: Sein Sinn und seine Bedeutung.* Reprint of the original 1922 publication. Erich Degreif Verlag, Oberteuringen, 134 pp.

Otto, H.-J. (1993). Waldbau in Europa - seine Schwächen und Vorzüge - in historischer Perspektive. *Forst und Holz,* Vol. 48, pp. 235-237.

Schütz, J-P. (1990). Heutige Bedeutung und Charakterisierung des naturnahen Waldbaus. *Schweizerische Zeitschrift für Forstwesen,* Vol. 141, pp. 609-614.

Skov- og Naturstyrelsen. (2002). *The Danish National Forest Programme in an International Perspective.* http://www.naturstyrelsen.dk/NR/rdonlyres/6BA78078-1188-494B-841E-EF89ECF0C064/13461/dnf_eng.pdf

Permissions

The contributors of this book come from diverse backgrounds, making this book a truly international effort. This book will bring forth new frontiers with its revolutionizing research information and detailed analysis of the nascent developments around the world.

We would like to thank Jorge Martín-García and Julio Javier Diez, for lending their expertise to make the book truly unique. They have played a crucial role in the development of this book. Without their invaluable contribution this book wouldn't have been possible. They have made vital efforts to compile up to date information on the varied aspects of this subject to make this book a valuable addition to the collection of many professionals and students.

This book was conceptualized with the vision of imparting up-to-date information and advanced data in this field. To ensure the same, a matchless editorial board was set up. Every individual on the board went through rigorous rounds of assessment to prove their worth. After which they invested a large part of their time researching and compiling the most relevant data for our readers. Conferences and sessions were held from time to time between the editorial board and the contributing authors to present the data in the most comprehensible form. The editorial team has worked tirelessly to provide valuable and valid information to help people across the globe.

Every chapter published in this book has been scrutinized by our experts. Their significance has been extensively debated. The topics covered herein carry significant findings which will fuel the growth of the discipline. They may even be implemented as practical applications or may be referred to as a beginning point for another development. Chapters in this book were first published by InTech; hereby published with permission under the Creative Commons Attribution License or equivalent.

The editorial board has been involved in producing this book since its inception. They have spent rigorous hours researching and exploring the diverse topics which have resulted in the successful publishing of this book. They have passed on their knowledge of decades through this book. To expedite this challenging task, the publisher supported the team at every step. A small team of assistant editors was also appointed to further simplify the editing procedure and attain best results for the readers.

Our editorial team has been hand-picked from every corner of the world. Their multi-ethnicity adds dynamic inputs to the discussions which result in innovative outcomes. These outcomes are then further discussed with the researchers and contributors who give their valuable feedback and opinion regarding the same. The feedback is then collaborated with the researches and they are edited in a comprehensive manner to aid the understanding of the subject.

Apart from the editorial board, the designing team has also invested a significant amount of their time in understanding the subject and creating the most relevant covers. They scrutinized every image to scout for the most suitable representation of the subject and create an appropriate cover for the book.

The publishing team has been involved in this book since its early stages. They were actively engaged in every process, be it collecting the data, connecting with the contributors or procuring relevant information. The team has been an ardent support to the editorial, designing and production team. Their endless efforts to recruit the best for this project, has resulted in the accomplishment of this book. They are a veteran in the field of academics and their pool of knowledge is as vast as their experience in printing. Their expertise and guidance has proved useful at every step. Their uncompromising quality standards have made this book an exceptional effort. Their encouragement from time to time has been an inspiration for everyone.

The publisher and the editorial board hope that this book will prove to be a valuable piece of knowledge for researchers, students, practitioners and scholars across the globe.

List of Contributors

Jorge Martín-García
Sustainable Forest Management Research Institute, Un0iversity of Valladolid – INIA, Palencia, Spain
Forestry Engineering, University of Extremadura, Plasencia, Spain

Julio Javier Diez
Sustainable Forest Management Research Institute, University of Valladolid – INIA, Palencia, Spain

Stéphane Couturier
Laboratorio de Análisis Geo-Espacial, Instituto de Geografía, UNAM, Mexico

André Monteiro and Carlos Souza Jr
Amazon Institute of People and The Environment-Imazon, Brazil

Pablo M. López Bernal and Guillermo E. Defossé
Consejo Nacional de Investigaciones Científicas y Técnicas – CONICET, Argentina
Centro de Investigación y Extensión Forestal Andino Patagónico – CIEFAP, Argentina
Universidad Nacional de la Patagonia San Juan Bosco – UNPSJB, Argentina

Pamela C. Quinteros
Consejo Nacional de Investigaciones Científicas y Técnicas – CONICET, Argentina
Centro de Investigación y Extensión Forestal Andino Patagónico – CIEFAP, Argentina

José O. Bava
Centro de Investigación y Extensión Forestal Andino Patagónico – CIEFAP, Argentina
Universidad Nacional de la Patagonia San Juan Bosco – UNPSJB, Argentina

Tohru Nakajima
Laboratory of Global Forest Environmental Studies, Graduate School of Agricultural and Life Sciences, The University of Tokyo, Yayoi, Bunkyo-ku, Japan
Laboratory of Forest Management, Graduate School of Agricultural and Life Sciences, The University of Tokyo, Yayoi, Bunkyo-ku, Tokyo, Japan

Jožica Gričar
Department of Yield and Silviculture, Slovenian Forestry Institute, Slovenia

John Schelhas
Southern Research Station, USDA Forest Service, G.W. Carver Agricultural Experiment Station, Tuskegee University, Tuskegee, USA

Joseph Molnar
Department of Agricultural Economics and Rural Sociology, Alabama Agricultural Experiment Station, Auburn University, USA

Ludmila La Manna
Centro de Investigación y Extensión Forestal Andino Patagónico, Universidad Nacional de la Patagonia San Juan Bosco, CONICET, Argentina

Roman Gebauer, Jindřich Neruda, Radomír Ulrich and Milena Martinková
Mendel University in Brno, Brno, Czech Republic

Nicholas Clarke
Norwegian Forest and Landscape Institute, Norway

Alessandro Paletto and Isabella De Meo
Agricultural Research Council – Forest Monitoring and Planning Research Unit (CRA-MPF), Villazzano di Trento, Italy

Fabrizio Ferretti
Agricultural Research Council – Apennine Forestry Research Unit (CRA-SFA), Isernia, Italy

Paolo Cantiani
Agricultural Research Council – Research Centre for Forest Ecology and Silviculture (CRA-SEL), Arezzo, Italy

Marco Focacci
Land / Forestry Resources Consultant, Sesto Fiorentino, Italy

Jørgen Bo Larsen
Forest & Landscape, University of Copenhagen, Denmark

Printed in the USA
CPSIA information can be obtained
at www.ICGtesting.com
JSHW011431221024
72173JS00004B/756